# Dynamic Documents with R and knitr
## Second Edition

## Yihui Xie
RStudio, Inc.

CRC Press
Taylor & Francis Group
Boca Raton   London   New York

CRC Press is an imprint of the
Taylor & Francis Group, an **informa** business

A CHAPMAN & HALL BOOK

# Chapman & Hall/CRC
# The R Series

## Series Editors

**John M. Chambers**
Department of Statistics
Stanford University
Stanford, California, USA

**Torsten Hothorn**
Division of Biostatistics
University of Zurich
Switzerland

**Duncan Temple Lang**
Department of Statistics
University of California, Davis
Davis, California, USA

**Hadley Wickham**
RStudio
Boston, Massachusetts, USA

## Aims and Scope

This book series reflects the recent rapid growth in the development and application of R, the programming language and software environment for statistical computing and graphics. R is now widely used in academic research, education, and industry. It is constantly growing, with new versions of the core software released regularly and more than 6,000 packages available. It is difficult for the documentation to keep pace with the expansion of the software, and this vital book series provides a forum for the publication of books covering many aspects of the development and application of R.

The scope of the series is wide, covering three main threads:
- Applications of R to specific disciplines such as biology, epidemiology, genetics, engineering, finance, and the social sciences.
- Using R for the study of topics of statistical methodology, such as linear and mixed modeling, time series, Bayesian methods, and missing data.
- The development of R, including programming, building packages, and graphics.

The books will appeal to programmers and developers of R software, as well as applied statisticians and data analysts in many fields. The books will feature detailed worked examples and R code fully integrated into the text, ensuring their usefulness to researchers, practitioners and students.

# Published Titles

**Stated Preference Methods Using R**, *Hideo Aizaki, Tomoaki Nakatani, and Kazuo Sato*

**Using R for Numerical Analysis in Science and Engineering**, *Victor A. Bloomfield*

**Event History Analysis with R**, *Göran Broström*

**Computational Actuarial Science with R**, *Arthur Charpentier*

**Statistical Computing in C++ and R**, *Randall L. Eubank and Ana Kupresanin*

**Reproducible Research with R and RStudio, Second Edition**, *Christopher Gandrud*

**Introduction to Scientific Programming and Simulation Using R, Second Edition**, *Owen Jones, Robert Maillardet, and Andrew Robinson*

**Nonparametric Statistical Methods Using R**, *John Kloke and Joseph McKean*

**Displaying Time Series, Spatial, and Space-Time Data with R**, *Oscar Perpiñán Lamigueiro*

**Programming Graphical User Interfaces with R**, *Michael F. Lawrence and John Verzani*

**Analyzing Sensory Data with R**, *Sébastien Lê and Theirry Worch*

**Parallel Computing for Data Science: With Examples in R, C++ and CUDA**, *Norman Matloff*

**Analyzing Baseball Data with R**, *Max Marchi and Jim Albert*

**Growth Curve Analysis and Visualization Using R**, *Daniel Mirman*

**R Graphics, Second Edition**, *Paul Murrell*

**Data Science in R: A Case Studies Approach to Computational Reasoning and Problem Solving**, *Deborah Nolan and Duncan Temple Lang*

**Multiple Factor Analysis by Example Using R**, *Jérôme Pagès*

**Customer and Business Analytics: Applied Data Mining for Business Decision Making Using R**, *Daniel S. Putler and Robert E. Krider*

**Implementing Reproducible Research**, *Victoria Stodden, Friedrich Leisch, and Roger D. Peng*

**Graphical Data Analysis with R**, *Antony Unwin*

**Using R for Introductory Statistics, Second Edition**, *John Verzani*

**Advanced R**, *Hadley Wickham*

**Dynamic Documents with R and knitr, Second Edition**, *Yihui Xie*

CRC Press
Taylor & Francis Group
6000 Broken Sound Parkway NW, Suite 300
Boca Raton, FL 33487-2742

© 2016 by Taylor & Francis Group, LLC
CRC Press is an imprint of Taylor & Francis Group, an Informa business

No claim to original U.S. Government works

Printed on acid-free paper
Version Date: 20150602

International Standard Book Number-13: 978-1-4987-1696-3 (Paperback)

---
**Library of Congress Cataloging-in-Publication Data**
---

Xie, Yihui (Mathematician)
   Dynamic documents with R and knitr / Yihui Xie. -- Second edition.
      pages cm
   "A CRC title."
   Includes bibliographical references and index.
   ISBN 978-1-4987-1696-3 (pbk. : alk. paper) 1. Statistics--Data processing. 2. Technical reports. 3. Corporation reports. 4. Mathematical statistics--Computer programs. 5. Technical writing--Data processing. 6. Technical writing--Computer programs. 7. R (Computer program language) 8. Knitr. I. Title.

QA276.4.X54 2015
005.13'3--dc23                                            2015021160
---

**Visit the Taylor & Francis Web site at**
**http://www.taylorandfrancis.com**

**and the CRC Press Web site at**
**http://www.crcpress.com**

To my parents

Shaobai Xie *and* Guolan Xie

# Contents

# *Preface*

We import a dataset into a statistical software package, run a procedure to get all results, then copy and paste selected pieces into a typesetting program, add a few descriptions, and finish a report. This is a common practice in writing statistical reports. There are obvious dangers and disadvantages in this process.

1. It is error-prone due to too much manual work.

2. It requires lots of human effort to do tedious jobs such as copying results across documents.

3. The workflow is barely recordable especially when it involves GUI (Graphical User Interface) operations, therefore it is difficult to reproduce.

4. A tiny change of the data source in the future will require the author(s) to go through the same procedure again, which can take nearly the same amount of time and effort.

5. The analysis and writing are separate, so close attention has to be paid to the synchronization of the two parts.

In fact, a report can be generated dynamically from program code. Just like a software package has its source code, a dynamic document is the source code of a report. It is a combination of computer code and the corresponding narratives. When we compile the dynamic document, the program code in it is executed and replaced with the output; we get a final report by mixing the code output with the narratives. Because we only manage the source code, we are free of all the possible problems above. For example, we can change a single parameter in the source code, and get a different report on the fly.

In this book, *dynamic documents* refer to the kind of source documents containing both program code and narratives. Sometimes we may just call them *source documents* since "dynamic" may sound confusing and ambiguous to some people (it does not mean interactivity or animations). We also use the term *report* frequently throughout the book, which really means the output document that was compiled from a dynamic document.

## Who Should Read This Book

This book is written for both beginners and advanced users. The main goal is to make writing reports easier: the "report" here can range from student homework or project reports, exams, books, blogs, and Web pages to virtually any documents related to statistical graphics, computing, and data analysis.

For beginners, Chapters 1 to 8 should be enough for basic applications (which have already covered many features); for power users, Chapters 9 to 11 can be helpful for understanding the extensibility of the **knitr** package.

Familiarity with LATEX and HTML can be helpful, but is not required at all. Once you get the basic idea, you can write reports in simple languages such as Markdown, which should be fairly easy for beginners to learn. Unless otherwise noted, all features apply to all document formats, although we primarily use LATEX for examples.

We recommend that readers take a look at the website RPubs (`http://rpubs.com`), which contains a large number of user-contributed documents. Hopefully they are convincing enough to show that it is quick and easy to write dynamic documents.

## Software Information and Conventions

The main tools we introduce in this book are the R language (R Core Team, 2015) and the **knitr** package (Xie, 2015b), with which this book was written, but the language in the documents is not restricted to R; for example, we can also integrate Python, awk, and shell scripts, etc., into the reports. For document formats, we mainly use LATEX, HTML, and Markdown.

Both R and **knitr** are available on CRAN (Comprehensive R Archive Network) as free and open-source software. You may download them from any CRAN mirrors, such as `http://cran.rstudio.com`. You can find their version information for this book in the R session information below:

```
sessionInfo()

## R version 3.2.0 (2015-04-16)
## Platform: x86_64-pc-linux-gnu (64-bit)
```

```
## Running under: Ubuntu 14.04.2 LTS
##
## locale:
##  [1] LC_CTYPE=en_US.UTF-8
##  [2] LC_NUMERIC=C
##  [3] LC_TIME=en_US.UTF-8
##  [4] LC_COLLATE=en_US.UTF-8
##  [5] LC_MONETARY=en_US.UTF-8
##  [6] LC_MESSAGES=en_US.UTF-8
##  [7] LC_PAPER=en_US.UTF-8
##  [8] LC_NAME=C
##  [9] LC_ADDRESS=C
## [10] LC_TELEPHONE=C
## [11] LC_MEASUREMENT=en_US.UTF-8
## [12] LC_IDENTIFICATION=C
##
## attached base packages:
## [1] stats     graphics  grDevices utils     datasets
## [6] base
##
## other attached packages:
## [1] knitr_1.10
##
## loaded via a namespace (and not attached):
## [1] formatR_1.2   tools_3.2.0   highr_0.5
## [4] stringr_0.6.2 evaluate_0.7
```

The **knitr** package is thoroughly documented on the website http://yihui.name/knitr/, and the most important page is perhaps http://yihui.name/knitr/options, where you can find the complete reference for chunk options (Section 5.1.1). The development version is hosted on Github: https://github.com/yihui/knitr; you can always check out the latest development version, file issues/feature requests, or even participate in the development by forking the repository and making changes by yourself. There are plenty of examples in the repository https://github.com/yihui/knitr-examples, including both minimal and advanced examples. Karl Broman prepared a very nice minimal tutorial for **knitr** at http://kbroman.org/knitr_knutshell, which can be useful for beginners to learn **knitr** quickly. There is also a wiki page maintained by Frank Harrell et al. from the Department of Biostatistics, Vanderbilt University, which introduced several tricks and useful experience of using **knitr**: http://biostat.mc.vanderbilt.edu.

Unlike many other books on R, we do not add prompts to R source

code in this book, and we comment out the text output by two hashes ##
by default, as you can see from the R session information before. The
reason for this convention is explained in Chapter 6. Package names
are in bold text (e.g., **rpart**), function names in italic (e.g., *paste()*), inline
code is formatted in a typewriter font (e.g., `mean(1:10, trim = 0.1)`),
and filenames are in sans serif fonts (e.g., figure/foo.pdf).

## Structure of the Book

Chapter 1 is an overview of dynamic documents, introducing the idea
of literate programming; Chapter 2 explains why dynamic documents
are important to scientific research from the viewpoint of reproducible
research; Chapter 3 gives a first complete example that covers basic
concepts and what we can do with **knitr**; Chapter 4 introduces a few
common text editors that support **knitr**, so that it is easier to compile
reports from source documents; and Chapter 5 describes the syntax for
different document formats such as LaTeX, HTML, and Markdown.

Chapters 6 to 11 explain the core functionality of the package. Chapters 6 and 7 present how to control text and graphics output from **knitr**.
Chapter 8 talks about the caching mechanism that may significantly reduce the computation time. Chapter 9 shows how to reuse source code
by chunk references and organize child documents. Chapter 10 consists
of an advanced topic — chunk hooks, which make a literate programming document really programmable and extensible. Chapter 11 illustrates how to integrate other languages, such as Python and awk, etc.,
into one report in the **knitr** framework.

Chapter 12 introduces some useful tricks that make it easier to write
documents with **knitr**. Chapter 13 shows how to publish reports in a
variety of formats including PDF, HTML, and HTML5 slides. Chapter
14 focuses on R Markdown v2, which can be converted to a large variety of document formats, including those in Chapter 13. Chapter 15
covers a few significant applications. Chapter 16 introduces other tools
for dynamic report generation, such as Sweave, other R packages, and
software in other languages. Appendix A is a guide to some internal
structures of **knitr**, which may be helpful to other package developers.

The topics from Chapters 6 to 11 are parallel to each other. For example, if you want to know more about graphics output, you can skip
Chapter 6 and jump to Chapter 7 directly.

In all, we will show how to improve our efficiency in writing reports, fine tune every aspect of a report, and go from program output to publication-quality reports.

## What's New in the Second Edition

The major new content in the second edition of this book is Chapter 14, which is an introduction to R Markdown v2. Then there are a few new sections: 6.3 (how to generate tables), 6.4 (how to define custom printing methods for objects in code chunks), 11.2.2 (the C/Fortran engines), 11.2.4 (the Stan engine), 11.3 (how to run engines in a persistent session), and 15.2 (how to start a local server to serve dynamic documents). There are many minor updates here and there in the book as well.

The second edition also introduces several changes according to the changes in the **knitr** package (the first edition was based on **knitr** 1.3).

- The default value of the chunk option tidy was changed from TRUE to FALSE, i.e., code chunks will not be automatically reformatted by default (Section 6.2.2).

- Inline R expressions are evaluated without *try()*, i.e., if an error occurs during the inline evaluation, R will stop immediately.

- The global R option digits is no longer modified in **knitr**; its default value is 7, and you can set options(digits = 4) if you want the old behavior.

- The plot hook function takes the plot filename as its first argument (Section 5.3), instead of a vector of length two (basename and extension).

- The preferred way to stop **knitr** in case of errors is to set the chunk option error = FALSE instead of the package option stop_on_error, which has been deprecated (Section 6.2.4).

- Syntax highlighting is also available for other languages (Chapter 11) such as Shell scripts, awk, and Python, etc., if the Highlight package is installed (Section 11.2.7).

- For external code chunks (Section 9.2), the preferred chunk delimiter is ## ---- instead of ## @knitr now.

To keep track of the changes in **knitr**, you can see the release notes for each version at https://github.com/yihui/knitr/releases.

## Acknowledgments

First, I want to thank my wireless router, which was broken when I started writing the core chapters of the first edition of this book (in the boring winter of Ames). Besides, I also thank my wife for not giving me the Ethernet cable during that period.

This book would certainly not have been possible without the powerful R language, for which I thank the R core team and its contributors. The seminal work of Sweave (by Friedrich Leisch and R-core) is the most important source of inspiration of **knitr**. Some additional features were inspired by other R packages including **cacheSweave** (Roger Peng), **pgfSweave** (Cameron Bracken and Charlie Sharpsteen), **weaver** (Seth Falcon), **SweaveListingUtils** (Peter Ruckdeschel), **highlight** (Romain Francois), and **brew** (Jeffrey Horner). The initial design was based on Hadley Wickham's **decumar** package, and the evaluator is based on his **evaluate** package. Both LyX and RStudio quickly included support to **knitr** after it came out, which made it a lot easier to write source documents, and I'd like to thank their developers (especially Jean-Marc Lasgouttes, JJ Allaire, and Joe Cheng); similarly I thank the developers of other editors such as Emacs/ESS. I do not know how to describe John MacFarlane's Pandoc. It is magic. "Yes, we do support Word! Welcome to the world of reproducible research!"

The R/**knitr** user community is truly amazing. There has been a lot of feedback since the beginning of its development in late 2011. I still remember some users shouted it from the rooftops when I released the first beta version. I appreciate this kind of excitement. Thousands of questions and comments in the mailing list (https://groups. google.com/group/knitr) and on the website StackOverflow (http:// stackoverflow.com/tags/knitr/) made this package far more powerful than I imagined. The development repository is on Github, where I have received nearly 800 issues and more than 160 pull requests from many contributors, including Ramnath Vaidyanathan, Taiyun Wei, Kirill Müller, and JJ Allaire (https://github.com/yihui/knitr/pulls).

```
# to see a full list of contributors
packageDescription("knitr", fields = "Authors@R")
```

I thank my PhD advisors at Iowa State University, Di Cook and Heike Hofmann, for their open-mindedness and consistent support for my research in this "non-classical" area of statistics. I also thank RStudio (http://www.rstudio.com) for providing me the freedom to work on the second edition of this book.

Lastly, I thank the reviewers Frank Harrell, Douglas Bates, Carl Boettiger, Joshua Wiley, Scott Kostyshak, and Jim Robison-Cox for their valuable advice on improving the quality of this book (which is the first book of my career), and I'm grateful to my editor John Kimmel, without whom I would not have been able to publish my first book quickly.

Yihui Xie
Ames, Iowa

# About the Author

**Yihui Xie** (http://yihui.name) is currently a software engineer at RStudio (http://www.rstudio.com). He earned his PhD from the Department of Statistics, Iowa State University. He is interested in interactive statistical graphics and statistical computing. As an active R user, he has authored several R packages, such as **animation, knitr, formatR, fun, mime, highr, servr,** and **Rd2roxygen,** among which the **animation** package won the 2009 John M. Chambers Statistical Software Award (ASA), and the **knitr** package was awarded the "Honorable Mention" prize in the "Applications of R in Business Contest 2012" thanks to Revolution Analytics.

In 2006, he founded the "Capital of Statistics" (http://cos.name), which has grown into a large online community on statistics in China. He initiated the first Chinese R conference in 2008, and has been organizing R conferences in China since then. During his PhD training at Iowa State University, he won the Vince Sposito Statistical Computing Award (2011) and the Snedecor Award (2012) in the Department of Statistics.

# *List of Figures*

# List of Tables

# 1

## *Introduction*

The basic idea behind dynamic documents stems from *literate programming*, a programming paradigm conceived by Donald Knuth (Knuth, 1984). The original idea was mainly for writing software: mix the source code and documentation together; we can either extract the source code out (called *tangle*) or execute the code to get the compiled results (called *weave*). A dynamic document is not entirely different from a computer program: for a dynamic document, we need to run software packages to compile our ideas (often implemented as source code) into numeric or graphical output, and insert the output into our literal writings (like documentation).

We explain the idea with a trivial example: suppose we need to write the value of $2\pi$ into a report; of course, we can directly write the number 6.2832. Now, if I change my mind and I want $6\pi$ instead, I may have to find a calculator, erase the previous value, and write the new answer. Since it is extremely easy for the computer to calculate $6\pi$, why not leave this job to the computer completely and free oneself from this kind of manual work? What we need to do is to leave the source code in the document instead of a hard-coded value, and tell the computer how to find and execute the source code. Usually we use special markers for computer code in the source report; e.g., we can write

```
The correct answer is {{ 6 * pi }}.
```

in which {{ and }} is a pair of markers that tell the computer 6 * pi is the source code and should be executed. Note here pi ($\pi$) is a constant in R.

If you know a Web scripting language such as PHP (which can embed program code into HTML documents), this idea should look familiar. The above example shows the *inline* code output, which means source code is mixed inline with a sentence. The other type of output is the *chunk* output, which gives the results from a whole block of code. The chunk output has much more flexibility; for example, we can produce graphics and tables from a code chunk.

Figure 1.1 was dynamically created with a chunk of R code, which is printed below:

1

FIGURE 1.1: A simulation of Brownian motion for 100 steps: $x_1 = \epsilon_1$, $x_{i+1} = x_i + \epsilon_{i+1}$, $\epsilon_i \overset{iid}{\sim} N(0,1)$, $i = 1, 2, \cdots, 100$

```
set.seed(1213)   # for reproducibility of random numbers
x <- cumsum(rnorm(100))
plot(x, type = "l", ylab = "$x_{i+1}=x_i+\\epsilon_{i+1}$",
    xlab = "step")
```

If we were to do this by hand, we would have to open R, paste the code into the R console to draw the plot, save it as a PDF file, and insert it into a LaTeX document with \includegraphics{}. This is both tedious for the author and difficult to maintain — supposing we want to change the random seed in *set.seed()*, increase the number of steps, or use a scatterplot instead of a line graph, we will have to update both the source code and the output. In practice, the computing and analysis can be far more complicated than the toy example in Figure 1.1, and more manual work will be required accordingly.

The spirit of dynamic documents may best be described by the philosophy of the ESS project (Rossini et al., 2004) for the S language:

The source code is real.

Philosophy for using ESS[S]

Since the output can be produced by the source code, we can maintain the source code only. However, in most cases, the direct output from the source code alone does not constitute a report that is readable for a human. That is why we need the literate programming paradigm. In this paradigm, an author has two tasks:

1. write program code to do computing, and
2. write narratives to explain what is being done by the program code

The traditional approach to doing the second task is to write comments for the code, but comments are often limited in terms of expressing the full thoughts of the authors. Normally we write our ideas in a paper or a report instead of hundreds of lines of code comments.

---

Let us change our traditional attitude to the construction of programs: Instead of imagining that our main task is to instruct a computer what to do, let us concentrate rather on explaining to humans what we want the computer to do.

Donald E. Knuth
Literate Programming, 1984

---

Technically, literate programming involves three steps:

1. parse the source document and separate code from narratives
2. execute source code and return results
3. mix results from the source code with the original narratives

These steps can be implemented in software packages, so the authors do not need to take care of these technical details. Instead, we only control what the output should look like. There are many details that we can tune for a report (especially for reports related to data analysis), although the idea of literate programming seems to be simple. For example, data reports often include tables, and Table 1.1 is a table generated from the R code below using the *kable()* function in **knitr**:

```
library(knitr)
kable(head(mtcars[, 1:6]))
```

Think how easy it is to maintain two lines of R code compared to maintaining many lines of messy LATEX code!

Generating reports dynamically by integrating computer code with

TABLE 1.1: A subset of the `mtcars` dataset: the first 6 rows and 6 columns.

|                   | mpg  | cyl | disp | hp  | drat | wt    |
|-------------------|------|-----|------|-----|------|-------|
| Mazda RX4         | 21.0 | 6   | 160  | 110 | 3.90 | 2.620 |
| Mazda RX4 Wag     | 21.0 | 6   | 160  | 110 | 3.90 | 2.875 |
| Datsun 710        | 22.8 | 4   | 108  | 93  | 3.85 | 2.320 |
| Hornet 4 Drive    | 21.4 | 6   | 258  | 110 | 3.08 | 3.215 |
| Hornet Sportabout | 18.7 | 8   | 360  | 175 | 3.15 | 3.440 |
| Valiant           | 18.1 | 6   | 225  | 105 | 2.76 | 3.460 |

narratives is not only easier, but also closely related to reproducible research, which we will discuss in the next chapter.

# 2

## *Reproducible Research*

Results from scientific research have to be reproducible to be trustworthy. We do not want a finding to be merely due to an isolated occurrence, e.g., only one specific laboratory researcher can produce the results on one specific day, and nobody else can produce the same results under the same conditions.

Reproducible research (RR) is one possible by-product of dynamic documents, but dynamic documents do not absolutely guarantee RR. Because there is usually no human intervention when we generate a report dynamically, it is likely to be reproducible since it is relatively easy to prepare the same software and hardware environment, which is everything we need to reproduce the results. However, the meaning of reproducibility can be beyond reproducing one specific result or one particular report. As a trivial example, one might have done a Monte Carlo simulation with a certain random seed and got a good estimate of a parameter, but the result was actually due to a "lucky" random seed. Although we can strictly reproduce the estimate, it is not actually reproducible in the general sense. Similar problems exist in optimization algorithms, e.g., different starting values can lead to different roots of the same equation.

Anyway, dynamic report generation is still an important step toward RR. In this chapter, we discuss a selection of the RR literature and practices of RR.

## 2.1 Literature

The term reproducible research was first proposed by Jon Claerbout at Stanford University (Fomel and Claerbout, 2009). The idea is that the final product of research is not only the paper itself, but also the full computational environment used to produce the results in the paper such as the code and data necessary for reproduction of the results and building upon the research.

Similarly, Buckheit and Donoho (1995) pointed out the essence of the scholarship of an article as follows:

---

An article about computational science in a scientific publication is not the scholarship itself, it is merely advertising of the scholarship. The actual scholarship is the complete software development environment and the complete set of instructions which generated the figures.

D. Donoho
WaveLab and Reproducible Research

---

That was well said! Fortunately, journals have been moving in that direction as well. For example, Peng (2009) provided detailed instructions to authors on the criteria of reproducibility and how to submit materials for reproducing the paper in the *Biostatistics* journal.

At the technical level, RR is often related to literate programming (Knuth, 1984), a paradigm conceived by Donald Knuth to integrate computer code with software documentation in one document. However, early implementations like WEB (Knuth, 1983) and Noweb (Ramsey, 1994) were not directly suitable for data analysis and report generation. There are other tools on this path of documentation generation, such as **roxygen2** (Wickham et al., 2015), which is an R implementation of Doxygen (van Heesch, 2008). Sweave (Leisch, 2002) was among the first implementations for dealing with dynamic documents in R (Ihaka and Gentleman, 1996; R Core Team, 2015). There are still a number of challenges that were not solved by the existing tools; for example, Sweave is closely tied to LaTeX and hard to extend. The **knitr** package (Xie, 2015b) was built upon the ideas of previous tools with a framework redesign, enabling easy and fine control of many aspects of a report. We will introduce other tools in Chapter 16.

An overview of literate programming applied to statistical analysis can be found in Rossini (2002). Gentleman and Temple Lang (2004) introduced general concepts of literate programming documents for statistical analysis, with a discussion of the software architecture. Gentleman (2005) is a practical example based on Gentleman and Temple Lang (2004), using an R package **GolubRR** to distribute reproducible analysis. Baggerly et al. (2004) revealed several problems that may arise with the standard practice of publishing data analysis results, which can lead to false discoveries due to lack of details for reproducibility (even with datasets supplied). Instead of separating results from computing, we can put everything in one document (called a *compendium* in Gentleman and Temple Lang (2004)), including the computer code and narratives. When we compile this document, the computer code will be executed, giving us the results directly.

## 2.2 Good and Bad Practices

The key to keep in mind for RR is that other people should be able to reproduce our results, therefore we should try our best to make our computation *portable*. We discuss some good practices for RR below and explain why it can be bad not to follow them.

- Manage all source files under the same directory and use relative paths whenever possible: absolute paths can break reproducibility, e.g., a data file like C:/Users/john/foo.csv or /home/joe/foo.csv may only exist in one computer, and other people may not be able to read it since the absolute path is likely to be different in their hard disk. If we keep everything under the same directory, we can read a data file with `read.csv('foo.csv')` (if it is under the current working directory) or `read.csv('../data/foo.csv')` (go one level up and find the file under the data/ directory); when we disseminate the results, we can make an archive of the whole directory (e.g., as a zip package).

- Do not change the working directory after the computing has started: *setwd()* is the function in R to set the working directory, and it is not uncommon to see `setwd('C:/path/to/some/dir')` in user's code, which is bad because it is not only an absolute path, but also has a global effect on the rest of the source document. In that case, we have to keep in mind that all relative paths may need adjustments since the root directory has changed, and the software may write the output in an unexpected place (e.g., the figures are expected to be generated in the ./figures/ directory, but are actually written to ./data/figures/ instead if we `setwd('./data/')`). If we have to set the working directory at all, do it in the very beginning of an R session; most of the editors to be introduced in Chapter 4 follow this rule, and the working directory is set to the directory of the source document before **knitr** is called to compile documents. If it is unavoidable or makes it much more convenient for you to write code after setting a different working directory, you should restore the directory later; e.g.,

```
f <- function(...) {
    # stores current dir to a variable owd
    owd <- setwd("a/different/dir/")
    # restore working dir when the function exits
    on.exit(setwd(owd), add = TRUE)
    # now you can work under a/different/dir
    ...
}
```

- Compile the documents in a clean R session: existing R objects in the current R session may "contaminate" the results in the output. It is fine if we write a report by accumulating code chunks one by one and running them interactively to check the results, but in the end we should compile a report in the batch mode with a new R session so all the results are freshly generated from the code.

- Avoid the commands that require human interaction: human input can be highly unpredictable; e.g., we do not know for sure which file the user will choose if we pop up a dialog box asking the user to choose a data file. Instead of using functions like *file.choose()* to input a file to *read.table()*, we should write the filename explicitly; e.g., `read.table('a-specific-file.txt')`.

- Avoid environment variables for data analysis: while environment variables are often heavily used in programming for configuration purposes, it is ill-advised to use them in data analysis because they require additional instructions for users to set up, and humans can simply forget to do this. If there are any options to set up, do it inside the source document.

- Attach *sessionInfo()* (or **devtools**::*session_info()*) and instructions on how to compile this document: the session information makes a reader aware of the software environment, such as the version of R, the operating system, and add-on packages used. Sometimes it is not as simple as calling one single function to compile a document, and we have to make it clear how to compile it if additional steps are required; but it is better to provide the instructions in the form of a computer script; e.g., a shell script, a Makefile, or a batch file.

These practices are not necessarily restricted to the R language, although we used R for examples. The same rules also apply to other computing environments.

Note that literate programming tools often require users to compile the documents in batch mode, and it is good for reproducible research, but the batch mode can be cumbersome for exploratory data analysis. When we have not decided what to put in the final document, we may need to interact with the data and code frequently, and it is not worth compiling the whole document each time we update the code. This problem can be solved by a capable editor such as RStudio and Emacs/ESS, which are introduced in Chapter 4. In these editors, we can interact with the code and explore the data freely (e.g., send or write R code in an associated R session), and once we finish the coding work, we can compile the whole document in the batch mode to make sure all the code works in a clean R session.

## 2.3 Barriers

Despite all the advantages of RR, there are some practical barriers, and here is a non-exhaustive list:

- the data can be huge: for example, it is common that high energy physics and next-generation sequencing data in biology can produce tens of terabytes of data, and it is not trivial to archive the data with the reports and distribute them

- confidentiality of data: it may be prohibited to release the raw data with the report, especially when it is involved with human subjects due to the confidentiality issues

- software version and configuration: a report may be generated with an old version of a software package that is no longer available, or with a software package that compiles differently on different operating systems

- competition: one may choose not to release the code or data with the report due to the fact that potential competitors can easily get everything for free, whereas the original authors have invested a large amount of money and effort

We certainly should not expect all reports in the world to be publicly available and strictly reproducible, but it is better to share even mediocre or flawed code or problematic datasets than not to share anything at all. Instead of persuading people into RR by policies, we may try to create tools that make RR easier than cut-and-paste, and **knitr** is such an attempt. The success of RPubs (http://rpubs.com) is evidence that an easy tool can quickly promote RR, because users enjoy using it. Readers can find hundreds of reports contributed by users in the RPubs website. It is fairly common to see student homework and exercises there, and once the students are trained in this manner, we may expect more reproducible scientific research in the future.

# 3

## A First Look

The **knitr** package is a general-purpose literate programming engine —
it supports document formats including LATEX, HTML, and Markdown
(see Chapter 5), and programming languages such as R, Python, awk,
C++, and shell scripts (Chapter 11). Before we get started, we need to
install **knitr** in R. Then we will introduce the basic concepts with minimal examples. Finally, we will show how to generate reports quickly
from pure R scripts, which can be useful for beginners who do not know
anything about dynamic documents.

### 3.1 Setup

Since **knitr** is an R package, it can be installed from CRAN in the usual
way in R:

```r
install.packages("knitr", dependencies = TRUE)
```

Note here that `dependencies = TRUE` is optional, and will install all
packages that are not absolutely necessary but can enhance this package with some useful features. The development version is hosted on
Github: `https://github.com/yihui/knitr`, and you can always check
out the latest development version, which may not be stable but contains the latest bug fixes and new features. If you have any problems
with **knitr**, the first thing to check is its version:

```r
packageVersion("knitr")
# if not the latest version, run
update.packages()
```

If you choose LATEX as the typesetting tool, you may need to install
MiKTEX (Windows, `http://miktex.org/`), MacTEX (Mac OS, `http://
tug.org/mactex/`), or TEXLive (Linux, `http://tug.org/texlive/`). If

you are going to work with HTML or Markdown, nothing else needs to be installed, since the output will be Web pages, which you can view with a Web browser.

Once we have **knitr** installed, we can compile source documents using the function *knit()*, e.g.,

```
library(knitr)
knit("your-file.Rnw")
```

A *.Rnw file is usually a LATEX document with R code embedded in it, as we will see in the following section and Chapter 5, in which more types of documents will be introduced.

## 3.2   Minimal Examples

We use two minimal examples written in LATEX and Markdown, respectively, to illustrate the structure of dynamic documents. We do not discuss the syntax of LATEX or Markdown for the time being (see Chapter 5 instead). For the sake of simplicity, the cars dataset in base R is used to build a simple linear regression model. Type ?cars in R to see detailed documentation. Basically it has two variables, speed and distance:

```
str(cars)

## 'data.frame':    50 obs. of  2 variables:
##  $ speed: num  4 4 7 7 8 9 10 10 10 11 ...
##  $ dist : num  2 10 4 22 16 10 18 26 34 17 ...
```

### 3.2.1   An Example in LATEX

Figure 3.1 is a full example of R code embedded in LATEX; we call this kind of documents *Rnw documents* hereafter because their filename extension is Rnw by convention. If we save it as a file minimal.Rnw and run knit('minimal.Rnw') in R as described before, **knitr** will generate a LATEX output document named minimal.tex. For those who are familiar with LATEX, you can compile this document to PDF via pdflatex. Figure 3.2 is the PDF document compiled from the Rnw document.

What is essential here is how we embedded R code in LATEX. In an Rnw document, <<>>= marks the beginning of code chunks, and @ terminates a code chunk (this description is not rigorous but is often easier

```
\documentclass{article}
\begin{document}
\title{A Minimal Example}
\author{Yihui Xie}
\maketitle

We examine the relationship between speed and stopping
distance using a linear regression model:
$Y = \beta_0 + \beta_1 x + \epsilon$.

<<model, fig.width=4, fig.height=3, fig.align='center'>>=
par(mar = c(4, 4, 1, 1), mgp = c(2, 1, 0), cex = 0.8)
plot(cars, pch = 20, col = 'darkgray')
fit <- lm(dist ~ speed, data = cars)
abline(fit, lwd = 2)
@

The slope of a simple linear regression is
\Sexpr{coef(fit)[2]}.
\end{document}
```

FIGURE 3.1: The source of a minimal Rnw document: see output in Figure 3.2.

to understand); we have four lines of R code between the two markers in this example to draw a scatterplot, fit a linear model, and add a regression line to the scatterplot. The command \Sexpr{} is used to embed inline R code, e.g., coef(fit)[2] in this example. We can write chunk options for a code chunk between << and >>=; the chunk options in this example specified the plot size to be 4 by 3 inches (fig.width and fig.height), and plots should be aligned in the center (fig.align).

In this minimal example, we have most basic elements of a report:

1. title, author, and date
2. model description
3. data and computation
4. graphics
5. numeric results

All the output is generated dynamically from R. Even if the data has

# A Minimal Example

### Yihui Xie

### April 11, 2015

We examine the relationship between speed and stopping distance using a linear regression model: $Y = \beta_0 + \beta_1 x + \epsilon$.

```
par(mar = c(4, 4, 1, 1), mgp = c(2, 1, 0), cex = 0.8)
plot(cars, pch = 20, col = "darkgray")
fit <- lm(dist ~ speed, data = cars)
abline(fit, lwd = 2)
```

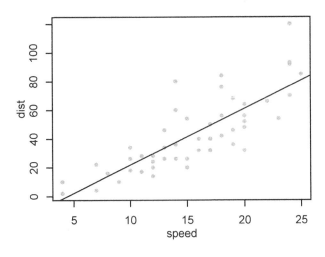

The slope of a simple linear regression is 3.9324088.

FIGURE 3.2: A minimal example in LaTeX with an R code chunk, a plot, and numeric output (regression coefficient).

```
---
title: A Minimal Example
---


We examine the relationship between speed and stopping
distance using a linear regression model:
$Y = \beta_0 + \beta_1 x + \epsilon$.

```{r fig.width=4, fig.height=3, fig.align='center'}
par(mar = c(4, 4, 1, 1), mgp = c(2, 1, 0), cex = 0.8)
plot(cars, pch = 20, col = 'darkgray')
fit <- lm(dist ~ speed, data = cars)
abline(fit, lwd = 2)
```


The slope of a simple linear regression is
`r coef(fit)[2]`.
```

FIGURE 3.3: The source of a minimal Rmd document: see output in Figure 3.4.

changed, we do not need to redo the report from the ground up, and the output will be updated accordingly if we update the data and recompile the report.

### 3.2.2 An Example in Markdown

LATEX may look overwhelming to beginners due to the large number of commands. By comparison, Markdown (Gruber, 2004) is a much simpler format. Figure 3.3 is a Markdown example doing the same analysis with the previous example:

The ideal output from Markdown is an HTML Web page, as shown in Figure 3.4 (in Mozilla Firefox). Similarly, we can see the syntax for R code in a Markdown document: ```{r} opens a code chunk, ``` terminates a chunk, and inline R code can be put inside `r `, where ` is a backtick.

A slightly longer example in **knitr** is a demo named notebook, which is based on Markdown. It shows not only the potential power of Markdown, but also the possibility of building Web applications with **knitr**. To watch the demo, run the code below:

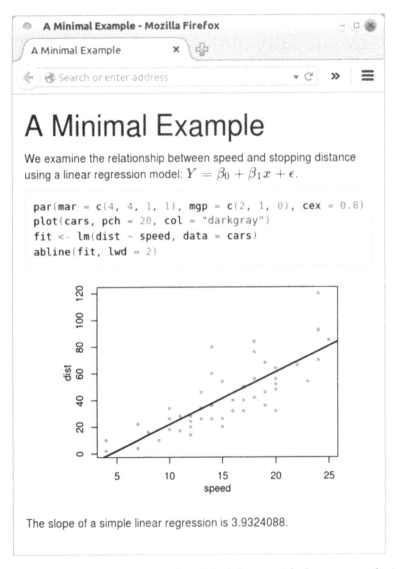

FIGURE 3.4: A minimal example in Markdown with the same analysis as in Figure 3.2, but the output is HTML instead of PDF now.

```
if (!require("shiny")) install.packages("shiny")
demo("notebook", package = "knitr")
```

Your default Web browser will be launched to show a Web note-book. The source code is in the left panel, and the live results are in the right panel. You are free to experiment with the source code and recompile the notebook.

## 3.3 Quick Reporting

If a user only has basic knowledge of R but knows nothing about **knitr**, or one does not want to write anything other than an R script, it is also possible to generate a quick report from this R script using the *stitch()* function.

The basic idea of *stitch()* is that **knitr** provides a template of the source document with some default settings, so that the user only needs to feed this template with an R script (as one code chunk); then **knitr** will compile the template to a report. Currently it has built-in templates for LaTeX, HTML, and Markdown. The usage is like this:

```
library(knitr)
stitch("your-script.R")
```

## 3.4 Extracting R Code

For a literate programming document, we can either compile it to a re-port (run the code), or extract the program code in it. They are called "weaving" and "tangling," respectively. Apparently the function *knit()* is for weaving, and the corresponding tangling function is *purl()* in **knitr**. For example,

```
library(knitr)
purl("your-file.Rnw")
purl("your-file.Rmd")
```

The result of tangling is an R script; in the above examples, the default output will be your-file.R, which consists of all code chunks in the source document.

So far we have been introducing the command line usage of **knitr**, and it is often tedious to type the commands repeatedly. In the next chapter, we show how a decent editor can help edit and compile the source document with one single mouse click or a keyboard shortcut.

# 4

## *Editors*

We can write documents for **knitr** with any text editor, because these documents are *plain text* files. For example, lightweight editors like Notepad under Windows or Gedit under Linux will work. The main reasons that we need special text editors are

1.  we want to input R code chunks more easily, e.g., input <<>>= and @ with a keyboard shortcut instead of typing these characters every time;

2.  we wish to call R and **knitr** to compile source documents to PDF/HTML within an editor instead of opening R and typing the command `knitr::knit()`, and even better, to send R code chunks to R from within the editor directly.

There are many mature and nice editors for LATEX, HTML, and Markdown documents, and some have integrated **knitr** within them, as we will explain in the following sections.

## 4.1  RStudio

RStudio is a relatively new editor specially targeted at R. It may be the best editor to start with for a beginner, since it has the most comprehensive support to Sweave and **knitr**. RStudio is cross-platform, free and open-source software; it is available at `http://www.rstudio.com`. Besides its excellent support for programming with R, it has a most notable feature that is missing in many other editors: it has a server version that looks identical to the desktop version, and we can use R in a Web browser after we have installed the server version on a Linux server.

The complete documentation can be found on the website. Here we only briefly introduce the features related to dynamic documents. If you are going to write Rnw documents (LATEX), the first thing to do to use **knitr** in RStudio is to change the option from the menu Tools ▷

19

FIGURE 4.1: Edit an Rnw document in RStudio: there is auto-completion inside the chunk header (we type "fig." and will see all candidates); the code chunk can be either inserted from the menu or a keyboard shortcut; the button Compile PDF supports one-click generation of PDF from Rnw.

Options ▷ Sweave; the default option for weaving (i.e., compiling) Rnw documents is Sweave, and we can switch it to knitr, as long as we have installed **knitr** in R. For more discussion about **knitr** vs. Sweave, see Section 16.1. If you plan to work with other types of documents such as R Markdown, you do not need to configure any options, and RStudio will give you tips to install the required packages if they are missing.

All document formats supported by RStudio can be found under the menu File ▷ New. Currently they include R Sweave, R Markdown, and R HTML. For all document formats, there is one-click compilation support, i.e., we can click a button to compile a source document to the corresponding output format (LaTeX to PDF, Markdown to HTML, and so on). We can input R code chunks with Ctrl + Alt + I; there is auto-completion of chunk options in the chunk header; e.g., if we type "fig." between << and >>= in an Rnw document, we will see possible candidates like `fig.width`, `fig.height`, and so on. The R code in chunks can be sent to the R console with Ctrl + Enter, just like what we do in a normal R script. In this way, we can run certain R code chunks interactively before we compile a whole document. Figure 4.1 is a screenshot of how an Rnw document looks in RStudio.

For an Rnw document, its final output format is usually PDF (via LaTeX). RStudio provides synchronization between the PDF document and the source document, which implies these features:

1. forward search: we can navigate from one line in the source document to an appropriate location in the PDF document that corresponds to the source line;

2. inverse search: we can also click in the PDF document and RStudio can bring us back to the corresponding lines in the Rnw source;

3. error navigation: when an error occurs in R or LaTeX, RStudio can bring us to a place in the source document that is the source of the error; this can help us fix problems in R or LaTeX code more quickly.

For R Markdown documents, RStudio provides one-click compilation to a variety of formats, including HTML. Besides, it can also base64 encode images and render LaTeX math expressions (through the MathJax library) in the HTML output. The former feature is to guarantee that the HTML page generated is self-contained, i.e., it does not depend on external images since they have been embedded in the page; the latter feature is especially useful for statisticians when they want to write math in a Web page.

The R Markdown (Rmd) format is fairly simple, and can be easily

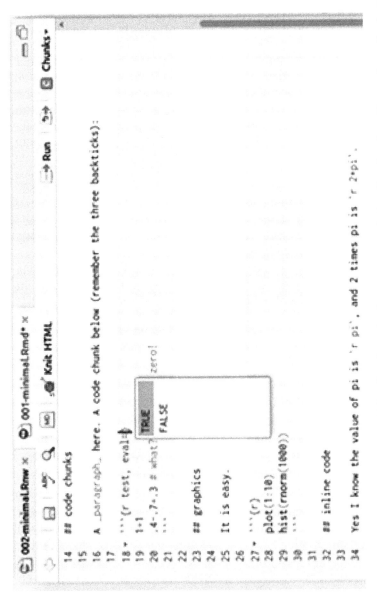

FIGURE 4.2: Edit an Rmd document in RStudio: there is also auto-completion for chunk option values; the button Knit HTML supports one-click generation of an HTML page from Rmd.

mastered in five minutes. Due to its simplicity, there has been a huge number of reports written in this format and published on RPubs, a free platform provided by RStudio to host **knitr** reports from users. See `http://rpubs.com` for more examples. Figure 4.2 shows a sample Rmd document in RStudio.

We mentioned quick reporting in Section 3.3, and this is also supported in RStudio. For an R script in RStudio, we can create an "R Notebook" (a report purely based on an R script) from it by clicking the button on the toolbar.

## 4.2  LyX

LyX is essentially a front-end for LaTeX, which has a nice GUI to assist document writing. On screen, it looks like many word processors, but at its core, it is LaTeX. One major difference between raw LaTeX editors and LyX is that we only see \alpha + \beta in raw LaTeX, whereas we see $\alpha + \beta$ in LyX, which is essentially \alpha + \beta behind the screen. Everything is LaTeX in LyX but our vision is not distorted by a full screen of backslashes.

Since version 2.0.3, LyX has started to support **knitr** as an official module. Details can be found at `http://yihui.name/knitr/demo/lyx/`. This module works in this way:

$$*.lyx \xrightarrow{LyX} *.Rnw \xrightarrow{R+knitr} \begin{cases} *.tex \xrightarrow{LaTeX} *.pdf & (weave) \\ *.R & (tangle) \end{cases}$$

Note that currently Rnw is the only possible format to use in LyX. It seems we are mixing R code with LyX, but LyX is really only a wrapper so we are actually embedding R code in Rnw documents.

For Linux and Mac OS users, the usage of the module is:

1. create a new LyX document;

2. go to Document ▷ Settings ▷ Modules and insert the module named Rnw (knitr);

3. insert R code chunks into the document with Insert ▷ TeX Code, then start typing <<>>= and @ as usual.

Click the View button on the toolbar or press Ctrl + R to compile the document to PDF and view the results. We can also extract R code from a LyX document from the menu File ▷ Export ▷ R/S code. A screenshot of LyX with R code is shown in Figure 4.3.

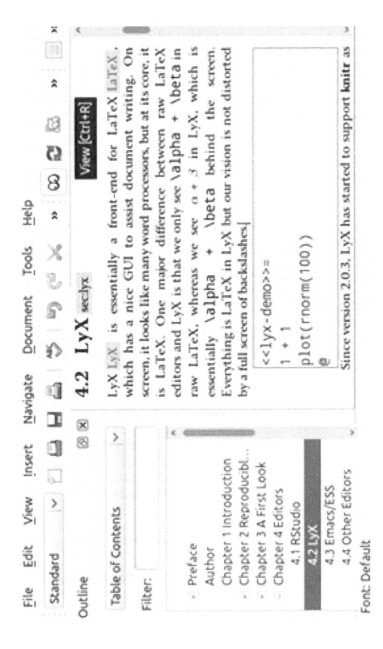

FIGURE 4.3: Using **knitr** in LyX: R code is inserted in a red box using the Rnw syntax; when we click the View button, we will see a PDF document compiled through **knitr** and LyX.

There is one more step before we can use the knitr module under Windows: go to Tools ▷ Preferences ▷ Paths ▷ PATH prefix and add the bin path of R there, which is often like C:\Program Files\R\R-x.x.x\bin and you can find it in R:

```
R.home("bin")
```

After you have made this change, you need to reconfigure LyX by Tools ▷ Reconfigure. This is to make sure LyX knows where R is installed so that it can call R and **knitr** to compile the Rnw document. Specifically, it needs to know where Rscript.exe is. If it is not present in PATH, the knitr module will be unavailable. This step is often not needed for Linux and Mac OS because these systems will put the R executable on PATH by default.

Although the graphical interface looks easy enough to use, we still strongly recommend users to master LaTeX before trying LyX; otherwise it can be difficult to diagnose LaTeX problems when errors occur. LyX is not Word, after all.

## 4.3 Emacs/ESS

ESS (Emacs Speaks Statistics) is an add-on package for the text editor Emacs (Rossini et al., 2004). It supports statistical software packages like R, S-Plus, SAS, JAGS, and so on. The support for **knitr** was added after version 12.09; before that, only Sweave was supported.

ESS is also free and open-source software; it is available at http://ess.r-project.org. After it has been installed along with Emacs, it is fairly easy to call **knitr** in Emacs. The default option for Rnw documents is Sweave, and we can change it to **knitr** with the following commands (in Emacs key notation, M stands for the Meta key, which is the Alt key on most keyboards, and M-x means to hold Meta and press x):

```
M-x customize-group
ess-R
```

Find the ess-swv-processor option and change it to knitr. Then we can create a new Rnw document, press M-n s to compile Rnw to TEX, and M-n P to compile TEX to PDF.

The support of Rmd documents and other document formats in ESS is still under development. According to the developers, this feature may come in ESS 13.03, and readers can pay attention to their official announcement in the future.

## 4.4 Other Editors

It is not hard to add support in other editors as long as they allow defining custom commands to compile documents. Generally speaking, the custom command looks like:

```
Rscript -e "library(knitr); knit('input.ext')"
```

This command calls R to load the **knitr** package and compile the input document named input.ext using the function *knit()*.

WinEdt (proprietary software) has a mode named R-Sweave to support **knitr**; and Tinn-R (free) has built-in support. It is also possible to configure other text editors such as Texmaker, Eclipse, TextMate, TEXShop, and Vim so that we can conveniently compile reports inside them. The configuration instructions are collected at http://yihui. name/knitr/demo/editors/.

# 5

## Document Formats

The design of the **knitr** package is flexible enough to process any plain text documents in theory. Below are the three key components of the design:

1. a source parser
2. a code evaluator
3. an output renderer

The parser parses the source document and identifies computer code chunks as well as inline code from the document; the evaluator executes the code and returns results; the renderer formats the results from computing in an appropriate format, which will finally be combined with the original documentation.

The code evaluator is independent of the document format, whereas the parser and the renderer have to take the document format into consideration. The former corresponds to the input syntax, and the latter is related to the output syntax.

## 5.1 Input Syntax

Regular expressions (Friedl, 2006, or see Wikipedia) are used to identify code blocks (chunks) and other elements such as inline code in a document. These regular expression patterns are stored in the `all_patterns` object in **knitr**. For example, the pattern for the beginning of a code chunk in an Rnw document is:

```
all_patterns$rnw$chunk.begin
```

```
## [1] "^\\s*<<(.*)>>=.*$"
```

In a regular expression, ^ means the beginning of a character string; \s* matches any number (including zero) of white spaces; .* matches any number of any characters. This regular expression means "any white spaces in the beginning of the line + << + any characters + >>=," therefore the lines below are possible chunk headers:

```
<<>>=
<<foo>>=
<<bar, echo=TRUE>>=
  <<a=1, b=2>>=
```

And these are not valid chunk headers (<< does not appear in the beginning of the line in the first one; there is only one > in the second one; = is missing in the third one):

```
hi<<>>=
<<foo>=
<<bar>>
```

Two more technical notes about the regular expression above:

1.  \s denotes a white space in regular expressions, but in R we have to write double backslashes because \\ in an R string really means one backslash (the first backslash acts as escaping the second character, which is also a backslash); the backslash as the *escape* character can be rather confusing to beginners, and the rule of thumb is, when you want a real backslash, you may need two backslashes;

2.  the braces () in the regular expression *group* a series of characters so that we can extract them with back references, e.g., we extract the second group of characters from abbbc:

    ```
    # [b]+ means to match 'b' for one or more times
    gsub("(a)([b]+)(c)", "\\2", "abbbc")
    ```

    ```
    ## [1] "bbb"
    ```

    We need to extract the chunk options in the chunk headers, and that is why we wrapped .* in () in the regular expression as <<(.*)>>=.

## 5.1.1   Chunk Options

As mentioned in Chapter 3, we can write chunk options in the chunk header. The syntax for chunk options is almost exactly the same as the syntax for function arguments in R. They are of the form

$$option = value$$

There is nothing to remember about this syntax due to the consistency with the syntax of R: as long as the option values are valid R code, they are valid to **knitr**. Besides constant values like echo = TRUE (a logical value) or out.width = '\\linewidth' (character string) or fig.height = 5 (a number), we can write arbitrary valid R code for chunk options, which makes a source document programmable. Here is a trivial example:

```
<<foo, eval=if (bar < 5) TRUE else FALSE>>=
```

Suppose bar is a numeric variable created in the source document before this chunk. We can pass an expression if (bar < 5) TRUE else FALSE to the option eval, which makes the option eval depend on the value of bar, and the consequence is we evaluate this chunk based on the value of bar (if it is greater than 5, the chunk will not be evaluated), i.e., we are able to *selectively* evaluate certain chunks. This example is supposed to show that we can write arbitrarily complicated R expressions in chunk options. In fact, it can be simplified to eval = bar < 5 since the expression bar < 5 normally returns TRUE or FALSE (unless bar is NA).

## 5.1.2 Chunk Label

The only possible exception is the chunk label, which does not have to follow the syntax rule. In other words, it can be invalid R code. This is due to both historical reasons (Sweave convention) and laziness (avoid typing quotes). Strictly speaking, the chunk label, as a part of chunk options, should take a character value, hence it should be quoted, but in most cases, **knitr** can take care of the unquoted labels and quote them internally, even if the "objects" used in the label expression do not exist. Below are all valid ways to write chunk labels:

```
<<foo>>=
<<foo-bar>>=
<<foo_bar>>=
<<"foo">>=
<<'foo-bar'>>=
<<label="foo">>=
<<echo=FALSE, label="foo-bar">>=
```

Chunk labels are supposed to be unique id's in a document, and they are mainly used to generate external files such as images (Chapter 7) and cache files (Chapter 8). If two non-empty chunks have the same label, **knitr** will stop and emit an error message, because there is potential danger that the files generated from one chunk may override the other chunk. If we leave a chunk label empty, **knitr** will automatically generate a label of the form unnamed-chunk-i, where i is an incremental chunk number from 1, 2, 3, ⋯.

### 5.1.3   Global Options

Chunk options control every aspect of a code chunk, as we will see in more detail in Chapters 6 through 11. If there are certain options that are used commonly for most chunks, we can set them as global chunk options using the object opts_chunk. Global options are shared across all the following chunks after the location in which the options are set, and local options in the chunk header can override global options. For example, we set the option echo to FALSE globally:

```
opts_chunk$set(echo = FALSE)
```

Then for the two chunks below, echo will be FALSE and TRUE, respectively:

```
<<foo>>=
1+1
@
<<bar, echo=TRUE>>=
rnorm(10)
@
```

### 5.1.4   Chunk Syntax

The original syntax of literate programming is actually this: use one marker to denote the beginning of computer code (<<>>=), and one marker to denote the beginning of the documentation (@). This has a subtle difference from what we introduced in Chapter 3. In the literate programming paradigm, this is what a source document may look like:

```
@
This is documentation.
@
```

```
Another line of documentation.
<<>>=
1 + 1 # some code
<<>>=
rnorm(10) # another code chunk
@
More documentation.
```

In **knitr** syntax, we open and close code chunks instead of opening code chunks and opening documentation chunks. The reason why we dropped the traditional syntax is that in a report, the code chunks often appear less frequently than normal text, so we only focus on the syntax for code chunks. It also looks more intuitive that we are "embedding" code into a report. Based on the new syntax, this is also a legitimate fragment of a source document for **knitr**:

```
Documentation here.
<<>>=
1+1
<<>>=
rnorm(10)
@
More documentation.
```

## 5.2   Document Formats

We have been using the syntax of Rnw documents as examples. Next we are going to introduce how to write R code in other document formats; Table 5.1 is a summary of the syntax. Note that code chunks can be indented by any number of spaces in all document formats.

### 5.2.1   Markdown

For an R Markdown (Rmd) document, we write code chunks between ```` ```{r} ```` and ```` ``` ````, and inline R code is written in `` `r ` ``. Chunk options are written before the closing brace in the chunk header. Note that the inline R code is not allowed to contain backticks, e.g., `` `r pi*2` `` is fine, but `` `r `pi`*2` `` is not; although `` `pi`*2 `` is valid R code, the parser is unable to know the first backtick is not for terminating the inline R code expression.

*Dynamic Documents with R and knitr*

TABLE 5.1: A syntax summary of all document formats: R LaTeX, R Markdown, R HTML, R reStructuredText, R AsciiDoc, R Textile, and brew.

| format | start | end | inline |
|---|---|---|---|
| Rnw | `<<*>>=` | `@` | `\Sexpr{x}` |
| Rmd | ` ```{r *} ` | ` ``` ` | `` `r x` `` |
| Rhtml | `<!--begin.rcode *` | `end.rcode-->` | `<!--rinline x-->` |
| Rrst | `.. {r *}` | `.. ..` | `:r:`x`` |
| Rtex | `% begin.rcode *` | `% end.rcode` | `\rinline{x}` |
| Rasciidoc | `// begin.rcode *` | `// end.rcode` | `` `r x` `` |
| Rtextile | `###. begin.rcode *` | `###. end.rcode` | `@r x@` |
| brew | | | `<% x %>` |

Markdown allows us to write using an easy-to-read, easy-to-write plain text format, then convert it to structurally valid XHTML or HTML. As long as one knows how to write emails, one can learn it in a few minutes: `http://en.wikipedia.org/wiki/Markdown`. Below is a short example:

```
# First level header

## Second level

This is a paragraph. This is **bold**, and _italic_.

- list item
- list item

Backticks produce the `<code>` tag. This is [a link](url),
and this is an ![image](url). A block of code (`<pre>` tag):

    1 + 1
    rnorm(10)

### Third level section title

You can write an ordered list:

1. item 1
2. item 2
```

The original Markdown syntax was designed to be simple, so it is inevitable to have some restrictions in terms of an authoring environment, such as the ability to write tables, LATEX math expressions, or, bibliography. In some cases, such as writing a short homework assignment, we do not need complicated features, so Markdown should work reasonably well.

One problem of Markdown is its derivatives: there are a number of variants such as Pandoc's Markdown (`http://johnmacfarlane.net/pandoc`), Github Flavored Markdown (`http://github.com`), kramdown (`http://kramdown.rubyforge.org`) and so on. These flavors may have their own definitions of how to write certain elements (such as tables). CommonMark (`http://commonmark.org`) is an effort at defining the Markdown syntax unambiguously, and Pandoc's Markdown is compatible with the CommonMark standards. Besides, Pandoc is probably the most comprehensive tool for Markdown at the moment. It added many useful extensions to the original Markdown such as:

1. Fenced code blocks within a pair of three backticks;

2. LATEX math via either plain LATEX (for PDF output) or MathJax (http://mathjax.org, for HTML output), which allows us to write math equations in Web pages using the LATEX syntax, i.e., $math$ or $$math$$;

3. Metadata for the document, e.g., the title, author, and date information;

4. Tables, with columns separated by white spaces or pipes;

5. Definition lists, footnotes, and citations, etc.

Below is how some of the extensions look:

```
---
title: The Title of My Report
author: Yihui Xie
---

Write code under ``` or indent by 4 spaces as usual.

```r
1 + 1
rnorm(10)
```

Inline math: $\alpha + \beta$. Display style:

$$f(x) = x^{2} + 1$$

A simple table from the citation [@joe2014]:

| id  | age | sex |
|:----|----:|:---:|
|  a  |  49 |  M  |
|  b  |  32 |  F  |
```

More importantly, Pandoc can convert Markdown to several other document formats, including PDF/LATEX, HTML, Word (Microsoft Word or OpenOffice), and presentation slides (either LATEX beamer or HTML5 slides). The R package **rmarkdown** (Allaire et al., 2015a) is based on **knitr** and Pandoc, and contains a few commonly used output formats so users can quickly create reasonably beautiful output by default.

The **rmarkdown** package was introduced by the RStudio developers, so it is not surprising that the R Markdown document format is best supported by RStudio. When we open or create an Rmd document in RStudio (File ▷ New ▷ R Markdown), we can see a wizard asking you which output format you want. We will cover R Markdown in detail in Chapter 14.

### 5.2.2 LaTeX

Markdown was primarily designed for the Web, and for more complicated typesetting purposes, LaTeX may be preferred. For example, this book was written in LaTeX. Oetiker et al. (1995) is a classic tutorial for beginners to learn LaTeX. The learning curve can be steep but it is rewarding if you care a lot about typesetting by yourself.

For LaTeX documents, R code chunks are embedded between <<>>= and @, and inline R code is written in \Sexpr{}, as we have seen many times before.

### 5.2.3 HTML

HTML (Hyper-Text Markup Language) is the language behind Web pages; normally we do not see HTML code directly because the Web browser has parsed it and rendered the elements. For example, when we see **bold** texts, the source code might be `<strong>bold</strong>`. Most Web browsers can show the HTML source code; e.g., for Firefox and Google Chrome, we can press Ctrl + U to view the page source.

There is a large (but limited) number of tags in HTML to represent different elements in a page. HTML is like LaTeX in the sense that we can have precise control over the typesetting by carefully organizing the tags/commands. The price to pay is that it may take a long time to write a document since there are many tags to type. That is why Markdown can be better for small-scale documents. Anyway, due to the fact that HTML has the full power, sometimes we have to use it. Below is an example of an HTML document:

```
<html>
<head>
  <title>This is an HTML page</title>
</head>
<body>
  <p>This is a <em>paragraph</em>.</p>
  <div>A <code>div</code> layer.</div>
  <!-- I'm a comment; you cannot see me. -->
</body>
</html>
```

To write R code in an HTML document, we use the comment syntax of HTML, e.g.,

```
<!--begin.rcode test-html, eval=TRUE
1 + 1
rnorm(10)
end.rcode-->

<p>And here is the value of pi: <!--rinline pi -->.</p>
```

### 5.2.4   reStructuredText

We can also embed R code in a reStructuredText (reST) document (`http://docutils.sourceforge.net/rst.html`), which is like Markdown but more powerful (and complicated accordingly). Below is an example of R code embedded in an R reST document:

```
A reST document for knitr
=========================

This is a reStructuredText document (*.Rrst). Here is how
we write R code for **knitr**:

.. {r test-rst, eval=TRUE}
1 + 1
rnorm(10)
.. ..

The value of pi is :r:`pi`.
```

The Docutils system (written in Python) is often used to convert reST documents to HTML.

### 5.2.5   AsciiDoc

AsciiDoc (`http://en.wikipedia.org/wiki/AsciiDoc`) is a plain-text document format that can be converted to multiple types of output, such as software documentation, articles, books, and HTML pages. Below is a minimal R AsciiDoc example for writing a book:

```
= The Book Title
:author: A Knitter
```

```
== The first chapter

Hello world!

// begin.rcode test, eval=TRUE
1 + 1
rnorm(10)
// end.rcode

The value of pi is `r pi`.
```

### 5.2.6  Textile

Textile is yet another lightweight markup language, and it is usually converted to HTML. You can find more information on the Wikipedia page http://en.wikipedia.org/wiki/Textile_(markup_language).
Here is an R Textile example demonstrating the syntax:

```
h1. Knitting Textile Files

Hello world!

###. begin.rcode test, tidy=FALSE
if (1 + 1 == 2) {
  'of course!'
}
###. end.rcode

And an inline expression @r 2*pi@.
```

### 5.2.7  Customization

It is possible to define one's own syntax to parse a source document. As we have seen before, the parsing is done through regular expressions. Internally, **knitr** uses the object knit_patterns to manage the regular expressions. For example, the three major patterns for this book are:

```
knit_patterns$get(
    c("chunk.begin", "chunk.end", "inline.code")
)
```

```
## $chunk.begin
## [1] "^\\s*<<(.*)>>=.*$"
##
## $chunk.end
## [1] "^\\s*@\\s*(%+.*|)$"
##
## $inline.code
## [1] "\\\\Sexpr\\{([^}]+)\\}"
```

To specify our own syntax, we can use `knit_patterns$set()`, which will override the default syntax, e.g.,

```
knit_patterns$set(
    chunk.begin = "^<<r(.*)", chunk.end = "^r>>$",
    inline.code = "\\{\\\\{([^}]+)\\}\\}"
)
```

Then we will be able to parse a document like this with the custom syntax:

```
<<r test-syntax, eval=TRUE
1 + 1
x <- rnorm(10)
r>>

The mean of x is {{mean(x)}}.
```

In practice, however, this kind of customization is often unnecessary. It is better to follow the default syntax, otherwise additional instructions will be required in order to compile a source document.

There is a series of functions with the prefix `pat_` in **knitr**, which are convenience functions to set up the syntax patterns, e.g., *pat_rnw()* calls `knit_hooks$set()` to set patterns for Rnw documents. All pattern functions include:

```
grep("^pat_", ls("package:knitr"), value = TRUE)

## [1] "pat_asciidoc" "pat_brew"     "pat_html"
## [4] "pat_md"       "pat_rnw"      "pat_rst"
## [7] "pat_tex"      "pat_textile"
```

When parsing a source document, **knitr** will first decide which pattern list to use according to the filename extension; e.g., *.Rmd documents use the R Markdown syntax. If the file extension is unknown,

**knitr** will further detect the code chunks in the document and see if the syntax matches with any existing pattern list; if it does, that pattern list will be used; e.g., for a file foo.txt, the extension txt is unknown to **knitr**, but if this file contains a code chunk that begins with ```{r}, **knitr** will use the R Markdown syntax automatically.

## 5.3   Output Renderers

The **evaluate** package (Wickham, 2015) is used to execute code chunks, and the eval() function in base R is used to execute inline R code. The latter is easy to understand and made possible by the power of "computing on the language" (R Core Team, 2014) of R. Suppose we have a code fragment 1+1 as a character string; we can parse and evaluate it as R code:

```r
eval(parse(text = "1+1"))

## [1] 2
```

For code chunks, it is more complicated. The **evaluate** package takes a piece of R source code, evaluates it, and returns a list containing results of six possible classes: character (normal text output), source (source code), warning, message, error, and recordedplot (plots).

In order to write these results into the output, we have to take the output format into consideration. For example, if the source code is 1+1 and the output format is TEX, we may use the verbatim environment, whereas if the output is supposed to be HTML, we may write <pre>1+1</pre> into the output instead. The key question is, how should we wrap up the raw results from R? This is answered by the knit_hooks object, which contains a list of output hook functions to construct the final output. A hook function is often defined in this form:

```r
hook_fun <- function(x, options) {
    # returns a character string with markup
}
```

In an output hook, x is usually the raw output from R, and options is a list of current chunk options. The hook names in knit_hooks corresponding to the output classes are listed in Table 5.2.

If we want to put the message output (emitted from *message()* func-

TABLE 5.2: Output hook functions and the object classes of results from the **evaluate** package.

| Class | Output hook | Arguments |
|---|---|---|
| source | source | x, options |
| character | output | x, options |
| recordedplot | plot | x, options |
| message | message | x, options |
| warning | warning | x, options |
| error | error | x, options |
|  | chunk | x, options |
|  | inline | x |
|  | text | x |
|  | document | x |

tion) into a custom LATEX environment, say, Rmessage, we can set the message hook as:

```
knit_hooks$set(message = function(x, options) {
    paste0("\\begin{Rmessage}\n", x, "\\end{Rmessage}")
})
```

Of course, we have to define the Rmessage environment in advance in the LATEX preamble, e.g.,

```
\newenvironment{Rmessage}{
  \rule[0.5ex]{1\columnwidth}{1pt} % a horizontal line
}{
  \rule[0.5ex]{1\columnwidth}{1pt}
}
```

Then, whenever we have a message in the output, we will see a horizontal line above and below it.

By default, **knitr** will set up a series of default output hooks for each output format, so normally we do not have to set up all the hooks by ourselves. A series of functions with the prefix render_ in **knitr** can be used to set up default output hooks for various output formats:

```
grep("^render_", ls("package:knitr"), value = TRUE)
## [1] "render_asciidoc"  "render_html"
## [3] "render_jekyll"    "render_latex"
## [5] "render_listings"  "render_markdown"
## [7] "render_rst"       "render_sweave"
## [9] "render_textile"
```

---

This is all you need to do if you want to go back to the Sweave style:
The quick brown fox jumps over the lazy dog the quick brown fox jumps over the lazy dog the quick brown fox jumps over the lazy dog.

```
> 1 + 1

[1] 2

> rnorm(30)

 [1] -0.56048 -0.23018  1.55871  0.07051  0.12929  1.71506  0.46092
 [8] -1.26506 -0.68685 -0.44566  1.22408  0.35981  0.40077  0.11068
[15] -0.55584  1.78691  0.49785 -1.96662  0.70136 -0.47279 -1.06782
[22] -0.21797 -1.02600 -0.72889 -0.62504 -1.68669  0.83779  0.15337
[29] -1.13814  1.25381
```

The quick brown fox jumps over the lazy dog the quick brown fox jumps over the lazy dog the quick brown fox jumps over the lazy dog.

---

FIGURE 5.1: The Sweave style in **knitr**: if we run *render_sweave()* in the beginning of an Rnw document, we will see the Sweave style.

The functions *render_latex()*, *render_html()*, and *render_markdown()* are called when the output formats are LaTeX, HTML, and Markdown, respectively; *render_sweave()* and *render_listings()* are two variants of LaTeX output — the former uses the traditional Sweave environments defined in Sweave.sty (e.g., Sinput and Soutput, etc.), and the latter uses the **listings** package in LaTeX to decorate the output. See Figure 5.1 and Figure 5.2 for how the two styles look.

Note that if we want to set up the output hooks, it is better to do it in the very beginning of a source document so that the rest of the output can be affected. For example, the chunk below can be the first chunk of an Rnw document (the chunk option include = FALSE means do not show anything from this chunk in the output because it is not interesting to the readers):

```
<<setup, include=FALSE>>=
render_sweave()
@
```

Then the output will be rendered in the Sweave style. This book used the default LaTeX style, which supports syntax highlighting, and code chunks are put in gray shaded boxes.

Among all output hooks in Table 5.2, there are five special hooks that need further explanation:

This is all you need to do if you want to use the listings package:
The quick brown fox jumps over the lazy dog the quick brown fox jumps
over the lazy dog the quick brown fox jumps over the lazy dog.

```
1 + 1
```

```
[1]  2
```

```
rnorm(30)
```

```
 [1] -0.56048 -0.23018  1.55871  0.07051  0.12929  1.71506  0.46092
 [8] -1.26506 -0.68685 -0.44566  1.22408  0.35981  0.40077  0.11068
[15] -0.55584  1.78691  0.49785 -1.96662  0.70136 -0.47279 -1.06782
[22] -0.21797 -1.02600 -0.72889 -0.62504 -1.68669  0.83779  0.15337
[29] -1.13814  1.25381
```

The quick brown fox jumps over the lazy dog the quick brown fox jumps
over the lazy dog the quick brown fox jumps over the lazy dog.

FIGURE 5.2: The listings style in **knitr**: *render_listings()* produces a style
like this (colored text and gray shading).

- the `plot` hook takes the filename as input x which is a character string
  of the filename (e.g., foo.pdf); below is a simplified version of the `plot`
  hook for LaTeX output (the actual hook is much more complicated than
  this, because there are many chunk options to take into account, such
  as `out.width` and `out.height`, etc.)

```
knit_hooks$set(plot = function(x, options) {
    paste("\\includegraphics{", x, "}", sep = "")
})
```

- the `chunk` hook takes the output of the whole chunk as input, which
  is generated from other hooks such as `source`, `output`, and `message`,
  etc.; for example, if we want to put the chunk output in a `div` tag with
  the class Rchunk in HTML, we can define the `chunk` hook as:

```
knit_hooks$set(chunk = function(x, options) {
    paste("<div class='Rchunk'>", x, "</div>")
})
```

then we need to define the style of Rchunk in the CSS stylesheet for
this HTML document;

- the `inline` hook is not associated with a code chunk; it defines how
  to format the output from inline R code. For example, we may want

to round all the numbers from inline output to 2 digits and we can define the `inline` hook as:

```
knit_hooks$set(inline = function(x) {
    if (is.numeric(x))
        x <- round(x, 2)
    as.character(x)  # convert x to character and return
})
```

**knitr** takes care of rounding in the default inline hook (Section 6.1), so we do not really have to reset this hook;

- the `text` hook processes text chunks, i.e., narratives; for example, we set up a hook to trim the white spaces around the text chunks:

```
knit_hooks$set(text = function(x) {
    gsub("^\\s*|\\s*$", "", x)
})
```

- the `document` hook is similar to the chunk hook, and it takes the output of the whole document as input x; this hook can be useful for post-processing the document; in fact, this book used this hook to add a vertical space \medskip{} under all table captions (before the `tabular` environment):

```
knit_hooks$set(document = function(x) {
    gsub("\\begin{tabular}", "\\medskip{}\\begin{tabular}",
        x, fixed = TRUE)
})
```

## 5.4 R Scripts

There is a special source document format in **knitr**, which is essentially an R script with roxygen comments (for more on roxygen, see Wickham et al. (2015) and Appendix A.1). We know a normal R comment starts with #, and a roxygen comment has an apostrophe after #, e.g.,

```
#' this is a roxygen comment
##' me too
```

Sometimes we do not want to mix R code with normal text, but write text in comments, so that the whole document is a valid R script. The function *spin()* in **knitr** can deal with such R scripts if the comments are written using the roxygen syntax. The basic idea of *spin()* is also inspired by literate programming: when we compile this R script, #' will be removed so that normal text is "restored," and R code will be evaluated. Anything that is not behind a roxygen comment is treated as a code chunk. To write chunk options, we can use another type of special comment #+ or #- followed by chunk options. Below is a simple example:

```
#' Introduce the method here; then write R code:
1 + 1
x <- rnorm(10)

#' It is also possible to write chunk options, e.g.,
#'
#+ test-label, fig.height=4
plot(x)
#' The document is done now.
```

We can save this script to a file called test.R, and compile it to a report:

```
library(knitr)
spin("test.R")
```

The *spin()* function has a format argument that specifies the output document format (default to R Markdown). For example, if format = 'Rnw', the R code will first be inserted between <<>>= and @, and then compiled to generate LATEX output.

This looks similar to the *stitch()* function in Section 3.3, which also creates a report based on an R script, but *spin()* makes it possible to write text chunks and *stitch()* can only use a predefined template, so there is less freedom.

# 6

## Text Output

From this chapter forward, we will start touching on the chunk options in **knitr**. First, in this chapter, we explain how to tune text output, including output from inline R code as well as text output from code chunks.

### 6.1 Inline Output

If the inline R code produces character results, they will be directly written into the output. When the result is numeric, scientific notation will be considered to denote the numbers that are too big or too small.

The threshold between scientific notation and fixed notation is the R option `scipen` (see `?options` for details). By default (`scipen = 0`), if a positive number is bigger than $10^4$ or smaller than $10^{-4}$ (this applies to the absolute values of negative numbers too), it will be denoted in scientific notation. Depending on the output format (LaTeX or HTML), **knitr** will use the appropriate code, such as `$3.14 \times 10^5$` or `3.14 &times; 10<sup>5</sup>`. The reason for scientific notation is to make it easier to read numbers such as small P-values, e.g., compare $0.000143$ with $1.43 \times 10^{-4}$.

Another R option `digits` controls how many digits a number should be rounded to. R's default is 7, which often makes a number unnecessarily "precise." We can change the defaults in the first chunk of a document, like:

```
# numbers >= 10^5 will be denoted in scientific
# notation, and rounded to 2 digits
options(scipen = 1, digits = 2)
```

For example, this book uses `digits = 4`, and a number 123456789 will become $1.2346 \times 10^8$ after the book source is compiled to PDF. Note that these two options are not specific to **knitr**; they are global options in R. If we are not satisfied with the default inline output, we can rewrite the `inline` hook as introduced in Section 5.3. Next we are going to introduce chunk options that affect the text output from code chunks.

For character results, we may have to take care of some special characters especially for LaTeX and HTML, e.g., % means comments in LaTeX, and a literal ampersand (&) has to be written as `&amp;` in HTML. See Section 12.3.6 for how to escape these characters if needed.

In most cases, characters are written as is in the output. For example, `\Sexpr{letters[1]}` produces "a" in the output of an Rnw document, and `` `r month.name[2]` `` in an Rmd document produces "February". A special case is the R HTML document: inline character results are written in the `<code></code>` tag by default, e.g., `<!--rinline 'ABC'-->` produces `<code class='knitr inline'>ABC</code>`. To get rid of the code tag, we can wrap the results in the function *I()*, which means to print the characters as is, e.g., `<!--rinline I('ABC')-->`.

## 6.2 Chunk Output

The "text output" in this section refers to any output from R that is not graphics, so even messages and warnings are classified as text output.

### 6.2.1 Chunk Evaluation

The chunk option `eval` (`TRUE` or `FALSE`) decides whether a code chunk should be evaluated. When a chunk is not evaluated, there will be no results returned except the original source code. This option can also take a numeric vector to specify which expressions are to be evaluated; in this case, the code that is set not to be evaluated will be commented out. For the chunk below, we set `eval = -2`, which means the second expression will not be evaluated:

```
1 + 1
```

```
## [1] 2
```

```
## if (TRUE) {
##     print("hi")
## }
dnorm(0)
```

```
## [1] 0.3989
```

## 6.2.2 Code Formatting

The function *tidy_source()* in the **formatR** package (Xie, 2015a) will be used to reformat R code (when the chunk option `tidy = TRUE`), e.g., it can add spaces and indentation, break long lines into shorter ones, and automatically replace the assignment operator = to <-; see the manual of **formatR** for details. The chunk option `tidy.opts` (a list) is passed to *tidy_source()* to control the formatting of R code. The example below shows the effect of `tidy = TRUE/FALSE`:

```
# option tidy=FALSE
for(k in 1:10){j=cos(sin(k)*k^2)+3;print(j-5)}
```

```
# option tidy=TRUE
for (k in 1:10) {
    j <- cos(sin(k) * k^2) + 3
    print(j - 5)
}
```

We can pass an argument `width.cutoff` to *tidy_source()* through the chunk option `tidy.opts = list(width.cutoff = 40)` so that the width of source code is roughly 40, e.g.,

```
0 + 1 + 2 + 3 + 4 + 5 + 6 + 7 + 8 + 9 + 0 +
    1 + 2 + 3 + 4 + 5 + 6 + 7 + 8 + 9 + 0 +
    1 + 2 + 3 + 4 + 5 + 6 + 7 + 8 + 9 + 0 +
    1 + 2 + 3 + 4 + 5 + 6 + 7 + 8 + 9
```

```
## [1] 180
```

```
# all arguments of tidy_source()
names(formals(formatR::tidy_source))
```

```
## [1] "source"      "comment"        "blank"
## [4] "arrow"       "brace.newline"  "indent"
## [7] "output"      "text"           "width.cutoff"
## [10] "..."
```

## 6.2.3 Code Decoration

Syntax highlighting comes by default in **knitr** (chunk option `highlight = TRUE`), since it enhances the readability of the source code — character strings, comments, and function names, etc., are in different colors.

This is achieved by the **highr** package (Qiu and Xie, 2015). This option only works for LaTeX and HTML output, and it is not necessary for Markdown because there are other libraries that can highlight code in Web pages; e.g., the JavaScript library **highlight.js** is widely used to do syntax highlighting for HTML pages.

For LaTeX output, the LaTeX package **framed** is used to decorate code chunks with a light gray background (as we can see in this book). If this package is not found in the system, a version will be copied directly from **knitr**. The output for HTML documents is styled with CSS, which looks similar to LaTeX (with gray shadings and syntax highlighting). The background color is controlled by the chunk option `background`, which takes a color value such as '#FF0000', 'red', or `rgb(1, 0, 0)` (as long as it is a valid color in R).

The prompt characters are removed by default because they mangle the R source code in the output and make it difficult to copy R code. The R output is masked in comments by default based on the same rationale (option `comment = '##'`). In fact, this was largely motivated from the author's experience of grading homework when he was a teaching assistant; with the default prompts, it is difficult to verify the results in the homework, because it is so inconvenient to copy the source code. Anyway, it is easy to revert to the output with prompts (set option `prompt = TRUE`), and we will quickly realize the inconvenience to the readers if they want to run the code in the output document, e.g., the chunk below uses `prompt = TRUE` and `comment = NA`:

```
> x <- rnorm(5)
> x
[1] -0.01156 -0.90915  0.37367  1.90694  0.16459
> var(x)
[1] 1.041
```

While this may seem to be irrelevant to reproducible research, we would argue that it is of great importance to design styles that look appealing and helpful at first glance, which can encourage users to write reports in this way.

For LaTeX output, we can also specify the font size of the chunk output via the `size` option, which takes the value of LaTeX font sizes such as footnotesize, small, large, and Large, etc., (the default size is normalsize). It is helpful to set a smaller font size when the output is long and the space is limited, e.g., in beamer slides. The chunk below uses `size = 'footnotesize'`:

```
<<font-size, size='footnotesize'>>=
x <- rnorm(20, mean = 5, sd = 3)
x^2

## [1] 5.039 8.314 10.604 5.749 28.855 38.501 14.089
## [8] 10.535 16.023 94.736 32.549 33.854 37.890 54.440
## [15] 41.333 31.910 8.445 2.227 46.454 25.077

@
```

### 6.2.4  Show/Hide Output

We can show or hide different parts of the text output including the source code, normal text output, warnings, messages, errors, and the whole chunk. Below are the corresponding chunk options with default values in the braces:

**echo** (TRUE) whether to show the source code; it can also take a numeric vector like the eval option to select which expressions to show in the output, e.g., echo = 1:3 selects the first 3 expressions, and echo = -5 means do not show the 5th expression.

**results** ('markup') how to wrap up the normal text output that would have been printed in the R console if we had run the code in R; the default value means to mark up the results in special environments such as LaTeX environments or HTML div tags; other possible values are:

'asis' write the raw output from R to the output document without any markups, e.g., the source code cat('emphasize') can produce an italic text in HTML when results = 'asis'; this is very useful when we use R to produce raw elements for the output, e.g., tables using the LaTeX markup (Section 6.3);

'hold' hold the text output and write to the end of the chunk;

'hide' this option value hides the normal text output.

**warning/error/message** (TRUE) whether to show warnings, errors, and messages in the output; usually these three types of messages are produced by *warning()*, *stop()*, and *message()* in R.

**split** (FALSE) whether to redirect the chunk output to a separate file (the filename is determined by the chunk label); for LaTeX, \input{} will be used if split = TRUE to input the chunk output from the file; for HTML, the <iframe> tag will be used; other output formats will ignore this option.

**include** (TRUE) whether to include the chunk output in the document;
when it is FALSE, the whole chunk will be absent in the output, but
the code chunk will still be evaluated unless eval = FALSE.

Below is an example that shows results = 'asis' and three types of
messages:

```r
b <- coef(lm(dist ~ speed, data = cars))
# write out the regression equation
cat(sprintf("$dist = %.02f + %.02f speed$", b[1], b[2]))
```

$dist = -17.58 + 3.93speed$

```r
x <- dnorm(0, sd = -1)   # will produce a warning

## Warning in dnorm(0, sd = -1):  NaNs produced

y <- 1 + "a"   # not possible; error

## Error:  non-numeric argument to binary operator

message("hello world!")

## hello world!
```

If we did not use the results option, we will see the raw LaTeX code
instead of an equation in the output:

```r
cat(sprintf("$dist = %.02f + %.02f speed$", b[1], b[2]))

## $dist = -17.58 + 3.93 speed$
```

As we have introduced in Section 5.1, we can use opts_chunk to set
global chunk options. For instance, if we want to suppress all warnings
and messages in the whole document, then we can do this in the first
chunk of the document:

```r
knitr::opts_chunk$set(warning = FALSE, message = FALSE)
```

When warning = FALSE (or message = FALSE), warnings (or messages) will be printed in the R console instead of the report output. If
you really want to suppress them, you have to call the function *sup-
pressWarnings()* (or *suppressMessages()*) on the R expression, e.g.,

```
suppressWarnings(1:2 + 1:3)  # no more warnings

## [1] 2 4 4

suppressMessages(message("foo"))
```

It may be very surprising to **knitr** users that **knitr** does not stop on errors! As we can see from the previous example, 1 + 'a' should have stopped R because that is not a valid addition operation in R (a number + a string). The default behavior of **knitr** is to act as if the code were pasted into an R console: if you paste 1 + 'a' to the R console, you will see an error message, but that does not halt R — you can continue to type or paste more code. To completely stop **knitr** when errors occur, we have to set the chunk option error = FALSE:

```
knitr::opts_chunk$set(error = FALSE)
```

### 6.2.5 Collapse Output

Currently this feature applies to R Markdown only. If a code chunk has many short R expressions, and each expression prints some output, it will be disturbing to read the output because R each expression and output fragment occupies a separate visual block. In this case, you can collapse all code and output fragments into one block using the chunk option collapse = TRUE. Here is an example:

```
1 + 1

## [1] 2

2 + 3

## [1] 5

if (TRUE) 1:10

## [1]  1  2  3  4  5  6  7  8  9 10
```

This is what the default output looks like (i.e., when collapse = FALSE):

```
1 + 1
```

```
## [1] 2
```

```
2 + 3
```

```
## [1] 5
```

```
if (TRUE) 1:10
```

```
##  [1]  1  2  3  4  5  6  7  8  9 10
```

### 6.2.6   Trim Blank Lines

The chunk option `strip.white` (TRUE by default) can be used to strip blank lines at the beginning and end of a source code chunk. For example, the blank line at the end of this chunk will be removed by default:

```
1 + 1
# a blank line below
```

## 6.3   Tables

Tables are essentially text output, but the first edition of this book did not cover table generation for a number of reasons:

1.  this functionality is orthogonal to **knitr** — as long as we can find another package to create the table, **knitr** can easily show it in the output with the chunk option `results = 'asis'`; a few good examples include **xtable** (Dahl, 2014), **Hmisc** (Harrell, 2015), and **tables** (Murdoch, 2012);

2.  it can be very challenging and complicated to generate tables for different document formats and different types of R objects, and the author has not found a perfect solution yet;

3.  sometimes graphics can present the information better than tables, and it is much easier to make plots.

However, it seems there is still high demand on this particular feature, so we will expand this topic a little bit. For LATEX tables, the packages mentioned above should work well. For HTML tables, **xtable** and **R2HTML** (Lecoutre, 2014) can be used. Additionally, Table 1.1 is an example of *kable()*, a simple function provided in **knitr** for LATEX, HTML, and Markdown tables. More importantly, the *kable()* function is aware of the output format, and can automatically generate a table of the appropriate format, e.g., for the same data object, it generates a LATEX table in an Rnw document, a Markdown table in an Rmd document, and an HTML table in an R HTML document. Therefore, you do not need to remember which type of document you are in, and just call *kable()* in your code chunk. The code chunk below shows the source code of tables of different formats:

```
# define a function to print the table source
kable_source <- function(...) cat(kable(...), sep = "\n")
# an example data frame
d <- data.frame(a = 1:3, b = pi * (1:3), c = c("ab", "cd",
    "efg"))
# the second argument of kable() is the output format
kable_source(d, "latex")


\begin{tabular}{rrl}
a & b & c\\
\hline
1 & 3.142 & ab\\
\hline
2 & 6.283 & cd\\
\hline
3 & 9.425 & efg\\
\end{tabular}

kable_source(d, "markdown")

|  a|      b|c   |
|--:|------:|:---|
|  1| 3.142|ab  |
|  2| 6.283|cd  |
|  3| 9.425|efg |
```

```
# center the first and third columns, and right align
# the second
kable_source(d, "markdown", align = c("c", "r", "c"))

| a |     b| c   |
|:-:|-----:|:---:|
| 1 | 3.142| ab  |
| 2 | 6.283| cd  |
| 3 | 9.425| efg |

kable_source(d, "pandoc")

  a     b   c
 ---  ------ ----
  1   3.142  ab
  2   6.283  cd
  3   9.425  efg

# use two digits
kable_source(d, "pandoc", digits = 2)

  a     b   c
 ---  ------ ----
  1   3.14  ab
  2   6.28  cd
  3   9.42  efg

# use different column names use two digits
kable_source(d, "pandoc", col.names = c("AAA", "BBB", "CCC"))

 AAA    BBB  CCC
 ----  ------ ----
  1   3.142  ab
  2   6.283  cd
  3   9.425  efg

kable_source(d, "html")

<table>
 <thead>
  <tr>
   <th style="text-align:right;"> a </th>
   <th style="text-align:right;"> b </th>
   <th style="text-align:left;"> c </th>
  </tr>
```

```
 </thead>
<tbody>
  <tr>
   <td style="text-align:right;"> 1 </td>
   <td style="text-align:right;"> 3.142 </td>
   <td style="text-align:left;"> ab </td>
  </tr>
  <tr>
   <td style="text-align:right;"> 2 </td>
   <td style="text-align:right;"> 6.283 </td>
   <td style="text-align:left;"> cd </td>
  </tr>
  <tr>
   <td style="text-align:right;"> 3 </td>
   <td style="text-align:right;"> 9.425 </td>
   <td style="text-align:left;"> efg </td>
  </tr>
</tbody>
</table>
```

If you simply want to display rectangular data as plain tables, *kable()* can be a good choice. If you want more advanced and complicated features such as conditional formatting (e.g., color certain rows/cells), you are advised to use other packages.

## 6.4 Automatic Printing

Under the hood, **knitr** uses the S3 generic function *knit_print()* to print objects in R code chunks by default. All visible objects are passed to *knit_print()* to render text output. Basically, *knit_print()* is the same as *print()*, but you can extend this S3 generic function by writing S3 methods for it without changing R's *print()* function. To know more details about this, please see the package vignette:

```
vignette("knit_print", package = "knitr")
```

The **printr** package (Xie, 2014) has provided several S3 methods for the *knit_print()* function. Once this package is loaded, you can just type the object names in a code chunk, and **knitr** will know how to print them automatically according to the output format. For example, when you type `??sunflower` in the R console (`??` means *help.search()* in R), you will see a help window pop up showing the search results using the keyword "sunflower." However, if you type this in an R code chunk, and compile it using **knitr**, normally you will see nothing because we cannot embed a transient help window in the output. Since `??` is essentially an R function that returns a special object of the class hsearch, the **printr** package has defined an S3 method *knit_print.hsearch()* to process the object of search results, so you can use the `??` command after loading the **printr** package:

```
library(printr)
??sunflower
```

| Package | Topic | Title |
|---|---|---|
| graphics | sunflowerplot | Produce a Sunflower Scatter Plot |
| grDevices | xyTable | Multiplicities of (x,y) Points, e.g., ... |

```
head(iris[, 1:4])
```

| Sepal.Length | Sepal.Width | Petal.Length | Petal.Width |
|---|---|---|---|
| 5.1 | 3.5 | 1.4 | 0.2 |
| 4.9 | 3.0 | 1.4 | 0.2 |
| 4.7 | 3.2 | 1.3 | 0.2 |
| 4.6 | 3.1 | 1.5 | 0.2 |
| 5.0 | 3.6 | 1.4 | 0.2 |
| 5.4 | 3.9 | 1.7 | 0.4 |

From the reader's perspective, this is cleaner than an explicit call to table-generating functions such as *kable()* in code chunks: the reader does not need to know what the table function was behind the scenes, and perhaps does not care either.

In fact, you do not have to use the *knit_print()* function. It is just the default value for the chunk option `render`, which takes a printing function. You are free to define another printing function and assign it to the `render` option. As a trivial example, you can use `render = print` to restore to the default printing behavior in the R console (*print()* is a function in base R).

## 6.5   Themes

The syntax highlighting theme can be adjusted or completely customized. If the default theme is not satisfactory, we can use the object `knit_theme` to change it. There are about 80 themes shipped with **knitr**, and we can view their names by `knit_theme$get()`. Here are the first 20:

```
head(knit_theme$get(), 20)
```

```
##  [1] "acid"        "aiseered"    "andes"
##  [4] "anotherdark" "autumn"      "baycomb"
##  [7] "bclear"      "biogoo"      "bipolar"
## [10] "blacknblue"  "bluegreen"   "breeze"
## [13] "bright"      "camo"        "candy"
## [16] "clarity"     "dante"       "darkblue"
## [19] "darkbone"    "darkness"
```

We can use `knit_theme$set()` to set the theme, e.g.,

```
knit_theme$set("autumn")
```

Each theme contains a set of color and font definitions, which will be translated to LATEX commands or CSS definitions (for HTML) in the end. Note that syntax highlighting themes only work for LATEX and HTML output. For Markdown, the **highlight.js** library also allows customization but that is beyond the scope of R and **knitr**. See http://bit.ly/knitr-themes for a preview of all these themes.

In the next chapter, we show how to control the graphics output.

# 7

# *Graphics*

Graphics are an important part of reports, and a lot of efforts have been made in **knitr** to make sure graphics output is natural and flexible. For example, **knitr** tries to mimic the behavior of the R console, and **grid** graphics (Murrell, 2011) may not need to be explicitly printed as long as the same code can produce plots in the R console (in some cases, however, they have to be printed, e.g., in a loop, because we have to do so in an R console); below is a chunk of code that will produce a plot in both the R console and **knitr** (see Figure 7.1):

```
library(ggplot2)
p <- qplot(carat, price, data = diamonds) + geom_hex()
p  # no need to print(p)
```

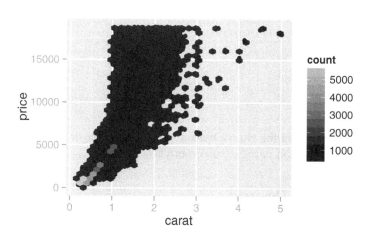

FIGURE 7.1: A plot created in **ggplot2** that does not need to be printed explicitly (by comparison, we have to print(p) in Sweave, which is very confusing; see Section 16.1).

## 7.1    Graphical Devices

There are more than 20 graphical devices supported in **knitr** through
the chunk option dev. For instance, dev = 'png' will use the *png()* device in the **grDevices** package in base R, and dev = 'CairoJPEG' uses
the *CairoJPEG()* device in the add-on package **Cairo** (it has to be installed first, of course). Here are the possible values for dev:

```
 [1] "bmp"          "postscript"   "pdf"
 [4] "png"          "svg"          "jpeg"
 [7] "pictex"       "tiff"         "win.metafile"
[10] "cairo_pdf"    "cairo_ps"     "quartz_pdf"
[13] "quartz_png"   "quartz_jpeg"  "quartz_tiff"
[16] "quartz_gif"   "quartz_psd"   "quartz_bmp"
[19] "CairoJPEG"    "CairoPNG"     "CairoPS"
[22] "CairoPDF"     "CairoSVG"     "CairoTIFF"
[25] "Cairo_pdf"    "Cairo_png"    "Cairo_ps"
[28] "Cairo_svg"    "tikz"
```

### 7.1.1    Custom Device

If none of these devices is satisfactory, we can provide the name of a
customized device function, which must be defined in this form before
it is used:

```
custom_dev <- function(file, width, height, ...) {
    # open the device here, e.g., pdf(file, width, height,
    # ...)
}
```

Then we can set the chunk option dev = 'custom_dev' (the device
name is the function name defined above).

### 7.1.2    Choose a Device

The default device for Rnw documents is PDF (*pdf()* in **grDevices**), and
for Rmd/Rhtml/Rrst documents, it is PNG (*png()* in **grDevices**), because normally PDF does not work in HTML output. The Cairo series
of devices can be very useful when we want high-quality raster images
such as PNG or JPEG, and the file sizes are often larger than the sizes

of plot files generated by *png()* or *jpeg()* in **grDevices**. The `CairoXXX` devices are from the **Cairo** package, and `Cairo_xxx` devices are from the **cairoDevice** package. The `quartz_xxx` devices are for Mac OS only.

For HTML output, we usually use raster images, but nowadays most Web browsers also support SVG as a format of vector graphics. One obvious advantage of vector graphics over raster graphics is their high quality, e.g., we can zoom in or zoom out a SVG image without loss of quality. We can use `dev = 'svg'` to generate SVG plots for Markdown or HTML. Again, the price to pay for the high quality is still the file size (this applies to R plots in general; SVG plots do not have to be larger than raster images, though).

Not all devices can be used for any output formats. As mentioned before, PDF does not automatically work in Web browsers at the moment; similarly, the `win.metafile` (Windows Metafile) device does not work with LaTeX.

### 7.1.3 Device Size

The chunk options `fig.width` and `fig.height` are passed to the graphical device to set the width and height of a plot (units in inches; default is 7 for both options), and the plot may be rescaled in the output using different options (Section 7.4). For bitmap devices such as *png()*, the default unit in R is pixel instead of inch, but **knitr** has made the units uniform to all devices. The chunk option `dpi` (dots per inch) is used to convert pixels to inches. It is 72 by default, meaning that 1 inch equals 72 pixels, so `fig.width = 7` means 504 pixels for PNG images.

### 7.1.4 More Device Options

Besides the options to set the size of plot files, we can pass even more arguments to the device via the `dev.args` option as a list. This is decided by the possible arguments of a specific graphical device. For example, we can pass `dev.args = list(pointsize = 10)` to the png device to change the pointsize, or `dev.args = list(family = 'Bookman')` to the pdf device to change the font family. Figure 7.2 was produced using the Bookman font family, although we cannot see the setting in the code below (it is in the source document):

```
plot(rep(0:1, 10), pch = 1:20, col = 2, xlab = "xlab font",
    ylab = "ylab font")
mtext("Bookman in the PDF device", side = 3, cex = 1.2)
text(6, 0.5, "Aa Bb Cc\nRr Ss Tt\nXx Yy Zz", cex = 1.5)
text(16, 0.5, "g", cex = 6, col = 3)
```

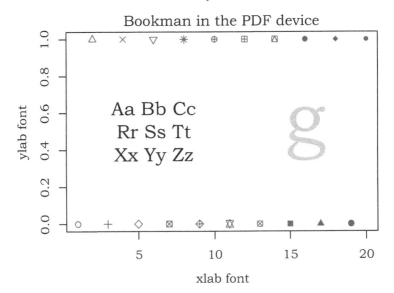

FIGURE 7.2: A plot using the Bookman font family: the chunk option for this plot is `dev.args = list(family = 'Bookman')` (with `dev = 'pdf'`).

We can compare the font family in Figure 7.2 with Figure 7.1, which used the default font family in the `pdf` device (Helvetica), and the two font styles are apparently different.

### 7.1.5 Encoding

For the `pdf` device, the options can be set globally via *pdf.options()*, i.e., the options set in this function will affect all `pdf` devices in the current R session. One important application of this function is to set the encoding for the `pdf` device in case of multi-byte characters in plots. For example, when we want to write the Euro sign or a letter A with the acute accent, we may need to set the encoding to CP1250 (to represent text in Central and Eastern European languages that use Latin script; see http://en.wikipedia.org/wiki/Windows-1250):

```
pdf.options(encoding = "CP1250")
```

For a complete list of possible encodings, see:

```
list.files(system.file("enc", package = "grDevices"))
```

```
##  [1] "AdobeStd.enc"   "AdobeSym.enc"   "CP1250.enc"
##  [4] "CP1251.enc"     "CP1253.enc"     "CP1257.enc"
##  [7] "Cyrillic.enc"   "Greek.enc"      "ISOLatin1.enc"
## [10] "ISOLatin2.enc"  "ISOLatin7.enc"  "ISOLatin9.enc"
## [13] "KOI8-R.enc"     "KOI8-U.enc"     "MacRoman.enc"
## [16] "PDFDoc.enc"     "TeXtext.enc"    "WinAnsi.enc"
```

Figure 7.3 shows a table of characters from the Windows-1250 code page, which is produced from the code below:

```
x <- c("\U20AC", "\U201A", "\U201E", "\U2026", "\U2020",
    "\U2021", "\U2030", "\U0160", "\U2039", "\U015A",
    "\U0164", "\U017D", "\U0179", "\U2018", "\U2019",
    "\U201C", "\U201D", "\U2022", "\U2013", "\U2014",
    "\U2122", "\U0161", "\U203A", "\U015B", "\U0165",
    "\U017E", "\U017A", "\U02C7", "\U02D8", "\U0141",
    "\U00A4", "\U0104", "\U00A6", "\U00A7", "\U00A8",
    "\U00A9", "\U015E", "\U00AB", "\U00AC", "\U00AE",
    "\U017B", "\U00B0", "\U00B1", "\U02DB", "\U0142",
    "\U00B4", "\U00B5", "\U00B6", "\U00B7", "\U00B8",
    "\U0105", "\U015F", "\U00BB", "\U013D", "\U02DD",
    "\U013E", "\U017C", "\U0154", "\U00C1", "\U00C2",
    "\U0102", "\U00C4", "\U0139", "\U0106", "\U00C7",
    "\U010C")
plot(c(1, 11), c(1, 6), type = "n", ann = F, axes = F)
box()
text(rep(1:11, 6), rep(1:6, each = 11), x)
```

If we do not set an appropriate encoding, we may see warnings like what appears below and the characters will be substituted by "..." (the character \U20AC below is the Euro sign €):

```
plot(1, main = "\U20AC")
```

```
## Warning: conversion failure on '€' in 'mbcsToSbcs': dot
substituted for <e2>
## Warning: conversion failure on '€' in 'mbcsToSbcs': dot
substituted for <82>
## Warning: conversion failure on '€' in 'mbcsToSbcs': dot
substituted for <ac>
```

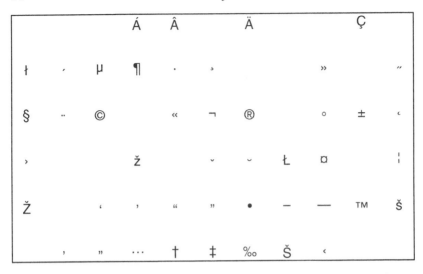

FIGURE 7.3: A table of the Windows-1250 code page: it only shows a subset of characters in the code page, such as the Euro sign and the letter A with an acute accent.

### 7.1.6   The Dingbats Font

According to the documentation of *pdf()*, the useDingbats argument can reduce the file size of PDF that contains small circles. If you use **knitr** in RStudio, this option is disabled by default. You may want to enable it by putting pdf.options(useDingbats = TRUE) in the source document if you have large scatterplots, and the PDF plot files will be smaller. Users with other editors do not need to take care of this option unless it is desired to set it to FALSE.

## 7.2   Plot Recording

All the plots in a code chunk are first recorded as R objects by the **evaluate** package and then "replayed" inside a graphical device to generate plot files. There are two sources of plots: first, whenever *plot.new()* or *grid.newpage()* is called (this happens before any R base and grid plot is created), **evaluate** will try to save a snapshot of the current plot if it exists; second, after each complete expression has been evaluated, a snapshot is also saved. For technical details, see ?setHook and ?recordPlot (both are functions in base R). To speed up recording, the null graphical device pdf (file = NULL) is used. Below is a simple example illustrating how a plot is recorded:

```
pdf(file = NULL)   # open a pdf device to record plots
## enable recording for the current device
dev.control("enable")
plot(rnorm(100))   # draw a plot
x <- recordPlot()
dev.off()

## pdf
##   2

str(x, 1)   # an R object of class recordedplot

## List of 3
##  $ :Dotted pair list of 8
##  $ : raw [1:35992] 00 00 00 00 ...
##  $ : NULL
##  - attr(*, "pid")= int 31856
##  - attr(*, "class")= chr "recordedplot"

print(x)   # redraw the plot object
```

The null device should work in most cases; one case in which it may not work is that where the plot contains multi-byte characters and it is complicated to deal with fonts (Murrell and Ripley, 2006). We can change the recording device by setting the device option in *options()*; for example, the *cairo_pdf()* device is better at dealing with non-standard fonts, and we can specify this device to record graphics instead:

```
options(device = function(width = 7, height = 7, ...) {
    cairo_pdf(tempfile(), width, height, ...)
})
```

Then we can also set the chunk option dev = 'cairo_pdf' to save plots as PDF files.

The **evaluate** package records plots per *expression* basis; in other words, the source code is split into individual complete expressions and **evaluate** will examine possible plot changes in snapshots after each single expression has been evaluated. Note that an R expression is not necessarily a line of code. For example, the code below consists of three expressions, out of which two are related to drawing plots (the first line

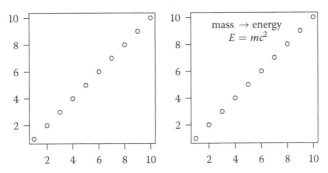

FIGURE 7.4: Three expressions produced two plots: the first expression does not draw any plots; the second draws a high-level plot; the third adds a low-level change (a text) to the plot. Section 7.6 will explain how the LATEX code was rendered in the right plot.

*par()* does not produce plots), therefore **evaluate** will produce two plots by default (see Figure 7.4):

```
par(mar = c(3, 3, 0.1, 0.1))
plot(1:10, ann = FALSE, las = 1)
if (TRUE) {
    text(5, 9, "mass $\\rightarrow$ energy\n$E=mc^2$")
}
```

This brings a significant difference with traditional tools in R for dynamic documents, since low-level plotting changes can also be recorded, whereas traditional tools (such as Sweave) do not capture these changes.

As a side note, there are high-level and low-level plotting commands in R: a high-level plotting command starts a new and complete plot (e.g., *plot()*, *hist()*, and *boxplot()*), and a low-level command often adds additional information to an existing plot (e.g., *text()*, *points()*, and *segments()*). It has to be called after a high-level plot has been created; see Murrell (2011) for more information.

Normally it is not straightforward, if not impossible, to capture low-level plotting changes as separate plots. The **evaluate** package has made this task easy.

Figure 7.5 shows two expressions producing two high-level plots. Recall that **knitr** tries to make graphics output natural — if we have two plots in a chunk, both will be shown in the output without any additional efforts.

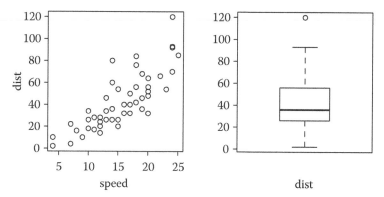

FIGURE 7.5: All high-level plots are captured and arranged side by side.

```
plot(cars)
boxplot(cars$dist, xlab = "dist")
```

The chunk option fig.keep controls which plots to keep in the output; fig.keep = 'all' means to keep low-level changes in separate plots; by default fig.keep = 'high', meaning that **knitr** will merge low-level plot changes into the previous high-level plot. This feature can be useful for teaching R graphics step by step; Figure 7.4 was one example, and Figure 7.6 (note it is one chunk instead of two) is another example of fig.keep = 'all' together with fig.show = 'asis' so that plots are put in the places where they were generated.

Note, however, low-level plotting commands inside another expression (a typical case is a loop) will not be recorded cumulatively, but high-level plotting commands, regardless of where they are, will always be recorded. For example, this chunk will only produce 2 plots instead of 21 plots because there are 2 complete expressions:

```
plot(0, 0, type = "n", ann = FALSE)
for (i in seq(0, pi, length = 20)) points(cos(i), sin(i))
```

But this will produce 20 plots as expected because *plot()* is a high-level plotting command even though there is only one expression:

```
for (i in seq(0, pi, length = 20)) {
    plot(cos(i), sin(i), xlim = c(-1, 1), ylim = c(-1, 1))
}
```

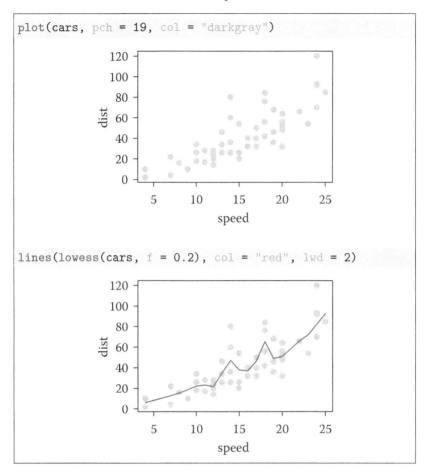

```
plot(cars, pch = 19, col = "darkgray")
```

```
lines(lowess(cars, f = 0.2), col = "red", lwd = 2)
```

FIGURE 7.6: Show plots right below the code: the option `fig.show` = 'asis' was used.

We can discard all previous plots and keep the last one only by `fig.keep` = 'last', or keep only the first plot by `fig.keep` = 'first', or discard all plots by `fig.keep` = 'none'. See Figure 7.7 for an example of keeping the last plot, and the code is below:

```
library(ggplot2)
pie <- ggplot(diamonds, aes(x = factor(1), fill = cut)) +
    xlab("cut") + geom_bar(width = 1)
pie + coord_polar(theta = "y")  # a pie chart
pie + coord_polar()  # the bullseye chart
```

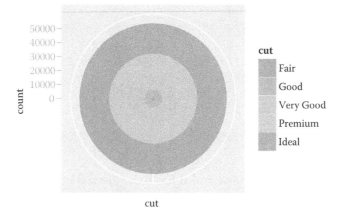

cut

FIGURE 7.7: Two plots were produced in this chunk, but only the last one was kept. This can be useful when we experiment with many plots, but only want the last result. (Adapted from the **ggplot2** website.)

A further note on plot recording: **knitr** examines all recorded plots (as R objects) and compares them sequentially; if the previous plot is a "subset" of the next plot (= previous plot + low-level changes), the previous plot will be removed by default (i.e., when fig.keep = 'high'). If two successive plots are identical, the second one will be removed by default, so it may be surprising that the following chunk will only produce one plot if we do not change the fig.keep option:

```
m <- matrix(1:100, ncol = 10)
image(m)
image(m * 2)  # exactly the same as previous plot
```

## 7.3 Plot Rearrangement

The chunk option fig.show determines whether to hold all plots in a chunk and "flush" all of them to the end of the chunk (fig.show = 'hold'; see Figures 7.4 and 7.5 for examples), or just insert them into the places where they were created (by default fig.show = 'asis'). Section 7.2 has shown an example of fig.show = 'asis' for two plots in one chunk.

```
<<clock-animation, fig.show='animate', interval=1>>=
par(mar = rep(3, 4))
for (i in seq(pi/2, -4/3 * pi, length = 12)) {
    plot(0, 0, pch = 20, ann = FALSE, axes = FALSE)
    arrows(0, 0, cos(i), sin(i))
    axis(1, 0, "VI"); axis(2, 0, "IX")
    axis(3, 0, "XII"); axis(4, 0, "III"); box()
}
@
```

FIGURE 7.8: A clock animation. It has to be viewed in Adobe Reader: click it to play/pause; there are also buttons to speed up or slow down the animation (the real animation is not shown here; see the graphics manual of **knitr** instead to see the real animation).

### 7.3.1 Animation

Beside 'hold' and 'asis', the option fig.show can take a third value: 'animate', which makes it possible to insert animations into the output document. In LaTeX, the package **animate** is used to put together image frames as an animation. For animations to work, there must be more than one plot produced in a chunk. The chunk option interval controls the time interval between animation frames; by default it is 1 second. Note we have to add \usepackage{animate} in the LaTeX preamble, because **knitr** does not add it automatically. Animations in the PDF output can only be viewed in Adobe Reader. There are animation examples in both the main manual and graphics manual of **knitr**, which can be found on the package website. Figure 7.8 shows the source code of a chunk that can produce an animation in a PDF document, but since animations will not work when printed on paper (of course), we did not show the output here.

For HTML output (including Markdown), this option also works, and there are three possible animation formats. The package option animation.fun can be used to set the hook function to generate animations. The **knitr** package has three built-in hook functions:

**hook_ffmpeg_html** Call FFmpeg to convert a series of image frames into a video file; the free software package FFmpeg has to be installed for this hook to work.

**hook_scianimator** Use the JavaScript library **SciAnimator** (https:// github.com/brentertz/scianimator) to display image frames one by one to form an animation; to use this hook, both jQuery and SciAnimator have to be included in the header of the HTML output, e.g.,

```
<head>
  <link rel="stylesheet" href="css/scianimator.css" />
  <script src="js/jquery-1.4.4.min.js"></script>
  <script src="js/jquery.scianimator.pack.js"></script>
</head>
```

These *.js and *.css files can be downloaded from the Github repository of SciAnimator; apparently this hook function requires fair knowledge of HTML.

**hook_r2swf** Use the **R2SWF** package (Qiu et al., 2015) to convert images to a Flash (SWF) animation; this hook only requires installation of the **R2SWF** package in R, and no additional software package or configurations are needed, so it may be the easiest one to use.

Here is how to set this package option:

```
opts_knit$set(animation.fun = hook_scianimator)
# or opts_knit$set(animation.fun = hook_r2swf)
```

### 7.3.2 Alignment

We can specify the figure alignment via the chunk option `fig.align` (possible values are `'default'`, `'left'`, `'center'`, and `'right'`). The global option for this book is `fig.align = 'center'` so most plots are centered. Figure 7.9 is an example of a right-aligned plot produced by the code chunk below:

```
stars(cbind(1:16, 10 * (16:1)), draw.segments = TRUE)
```

For LATEX, **knitr** uses the horizontal fill (\hfill{}) on the left or right of a plot to right- or left-align a plot, and {\centering } is used to center a plot. For HTML output, a CSS class is attached to a plot to align it, e.g., for a left-aligned plot, it is put in a div element <div class='rimage left'></div>, and the CSS definition for the left class is float: left;. The alignment option is ignored in Markdown.

FIGURE 7.9: A right-aligned plot adapted from ?stars: the chunk option is fig.align = 'right'.

## 7.4 Plot Size in Output

The fig.width and fig.height options specify the size of plots in the graphical device, and the real size in the output document can be different (specified by out.width and out.height). When there are multiple plots per code chunk, it is possible to arrange multiple plots side by side. For example, in LaTeX we only need to set out.width to be less than half of the current line width, e.g., out.width = '.49\\linewidth' (this is a common setting for plots in this chapter), and the plots will be inserted in the LaTeX document using the code as below:

```
\includegraphics[width=.49\linewidth]{plot-foo}
```

Note that fig.width and fig.height normally take numeric values, whereas out.width and out.height take character values that depend on the output format, e.g., out.width = '50%' (50% of the width of the parent container) or '480px' (480 pixels) for figures in HTML output.

The default value for out.width for LaTeX output is \maxwidth which is not a standard LaTeX length and was defined as:

```
% maxwidth is the original width if it's less than linewidth
% otherwise use linewidth
\makeatletter
\def\maxwidth{ %
  \ifdim\Gin@nat@width>\linewidth
    \linewidth
  \else
```

```
    \Gin@nat@width
  \fi
}
\makeatother
```

This is a reasonable default value because when the plot is wider than the line width, it will be resized to fit the line width; otherwise its original width is used. In other words, the plots will never exceed the page margin in LaTeX by default.

For Retina displays, the chunk option `fig.retina` can be used to improve the image quality in HTML output. For example, if you use `fig.retina` = 2, the actual size of the image will be twice as large as the size specified by `fig.width` and `fig.height`, but the display size will be half of the actual size, i.e., the size used for the display is still `fig.width` and `fig.height`.

## 7.5   Extra Output Options

The chunk option `out.extra` can be used to write more options to tune the plot output. For LaTeX output, this option is written inside the square brackets after `\includegraphics`, e.g., we can set `out.extra` = 'angle=90' to rotate a figure by 90 degrees; for HTML output, it is written in the `<img />` tag, e.g., use `out.extra` = 'style="display:none"' to hide a plot through the CSS attribute `display`.

The options `out.width`, `out.height`, and `out.extra` are recycled in the sense that if there are multiple plots in a chunk, these options will be first extended to the length of plots, and the *i*-th element of each option will be applied to the *i*-th plot. Figure 7.10 shows two plots in the same code chunk but with different angles for rotation (`out.extra` = `sprintf('angle=%d', c(-30, 90)')`).

```
plot(1:10, pch = 1:10, col = 1:10, cex = 2, lwd = 2)
lines(1:10, type = "h", col = "lightgray")
plot(rnorm(30), pch = 21, cex = 1.5, col = "darkgreen",
    bg = "lightgreen")
```

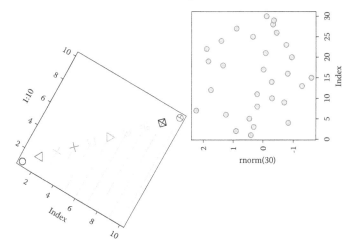

FIGURE 7.10: Rotate two plots with different angles: the first plot is rotated by -30 degrees, and the second is rotated by 90 degrees.

## 7.6 The *tikz()* Device

Beside PDF, PNG, and other traditional R graphical devices, **knitr** has special support for TikZ graphics (Tantau, 2008) via the **tikzDevice** package (Sharpsteen and Bracken, 2015), which is similar to the feature of the **pgfSweave** package. If we set the chunk option dev = 'tikz', the *tikz()* device in **tikzDevice** will be used to generate plots. A plot file created by the *tikz()* device is essentially a LaTeX file, although **knitr** uses the filename extension *.tikz.

Options sanitize (for escaping special TeX characters in plots such as \ and %) and external are related to the *tikz()* device: see the documentation of *tikz()* for details. Note that external = TRUE in **knitr** means standAlone = TRUE in *tikz()*, and the TikZ graphics output will be compiled to PDF *immediately* after it is created, so the "externalization" does not depend on the official but complicated externalization commands in the **tikz** package in LaTeX (see the manual for PGF and TikZ). The advantage of externalization is that it saves the time of compiling TikZ graphics to PDF when the main LaTeX document is compiled.

To maintain consistency in (font) styles, **knitr** will read the preamble of the input document and pass it to the *tikz()* device, so that the

$$p(\theta \mid \mathbf{x}) \propto \pi(\theta)f(\mathbf{x} \mid \theta)$$

FIGURE 7.11: The traditional approach to writing math expressions in plots: we have to carefully construct an R expression object.

$$p(\theta|\mathbf{x}) \propto \pi(\theta)f(\mathbf{x}|\theta)$$

FIGURE 7.12: Write math in native LaTeX with the *tikz()* device: everything is natural LaTeX code. The function *paste()* was used only for the sake of typesetting this book (break the long character string into two lines that could have been written in the same string).

font style in the plots will be the same as the style of the whole LaTeX document.

Besides consistency of font styles, the *tikz()* device also enables us to write arbitrary LaTeX expressions into R plots. A typical use is to write math expressions. The traditional approach in R is to use an *expression()* object to write math symbols in the plot, and for the *tikz()* device, we only need to write normal LaTeX code. Below is an example of a math expression $p(\theta|\mathbf{x}) \propto \pi(\theta)f(\mathbf{x}|\theta)$ using the two approaches respectively. This is a code chunk for Figure 7.11 (traditional approach):

```
plot(0, type = "n", ann = FALSE)
text(0, expression(p(theta ~ "|" ~ bold(x)) %prop%
    pi(theta) * f(bold(x) ~ "|" ~ theta)), cex = 2)
```

With the *tikz()* device, it is both straightforward (if we are familiar with LaTeX) and more beautiful (Figure 7.12):

```
plot(0, type = "n", ann = FALSE)
text(0, paste("$p(\\theta|\\mathbf{x})", "\\propto",
    "\\pi(\\theta)f(\\mathbf{x}|\\theta)$"), cex = 2)
```

Note that it is not impossible to improve the fonts in the traditional approach; see Murrell and Ripley (2006) for details.

One disadvantage of the *tikz()* device is that LaTeX may not be able to handle large tikz files (LaTeX can run out of memory). For example, an R plot with tens of thousands of graphical elements may fail to compile in LaTeX if we use the *tikz()* device. In such cases, we can switch to the PDF or PNG device, or reconsider our decision on the type of plots, e.g., a scatterplot with millions of points is usually difficult to read, and a contour plot or a hexagon plot showing the 2D density can be a better alternative (they are smaller in size).

When using XeTeX or LuaTeX instead of PDFTeX to compile the document, we need to set the `tikzDefaultEngine` option before all plot chunks (preferably in the first chunk):

```
options(tikzDefaultEngine = "xetex")   # or 'luatex'
```

This is useful and often necessary to compile tikz plots that contain multi-byte characters.

## 7.7   Figure Environment

For plots in LaTeX output, **knitr** can automatically create the `figure` environment. This happens when we set the `fig.cap` option to character strings of figure captions. A figure environment looks like this:

```
\begin{figure}[position]
  % e.g., \includegraphics{foo} here
  \caption[short caption]{full caption.} \label{label}
\end{figure}
```

The `fig.cap` option specifies the full caption. Other relevant chunk options are (default values in braces):

**fig.env** (`'figure'`) the environment name to use, e.g., we can use the `marginfigure` or `sidewaysfigure` environment instead of the default `figure` environment

**fig.pos** (") position arrangement of a figure, e.g., `'tbp'`

**fig.scap** (NULL) the short caption; if NULL, all the words before . or ; or : in `fig.cap` will be used as the short caption; if NA, it will be ignored

**fig.lp** (`'fig:'`) the label prefix; for each chunk, the figure label is derived from the chunk label, with `fig.lp` as the prefix, e.g., if the chunk label is `foo`, the figure label will be `fig:foo` by default; figure labels can be used to cross-reference figures with the LaTeX command `\ref{}`

If there are multiple plots produced from a chunk, we can create multiple figure environments accordingly. In this case, `fig.cap` has to be a vector of figure captions, and the length is equal to the number of plots;

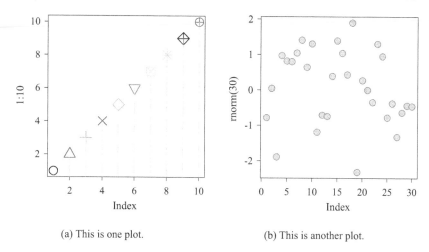

(a) This is one plot.            (b) This is another plot.

FIGURE 7.13: A figure environment with sub-figures: it can be created by the `fig.subcap` and `fig.cap` options.

meanwhile, the chunk option `fig.show` should be `'asis'` (otherwise only one figure environment will be created).

In the case of multiple plots per chunk, an alternative approach to arrange plots is to use sub-figures, which requires the **subfig** package in the LaTeX preamble. To put all plots in sub-figure environments, we need to assign sub-captions to plots via the `fig.subcap` option, e.g., `fig.subcap = c('sub caption 1', 'sub caption 2')`, and `fig.cap = 'full main caption.'` will generate a figure environment with sub-floats (`\subfloat{}`) in it like this:

```
\begin{figure}
\subfloat[sub caption 1\label{foo1}]{\includegraphics{foo1}}
\subfloat[sub caption 2\label{foo2}]{\includegraphics{foo2}}
\caption[short main caption]{full main caption.} \label{foo}
\end{figure}
```

Figure 7.13 shows two plots in one figure environment. The output width of plots was set to `.49\linewidth` so they can sit side by side.

Apparently the figure environment is specific to LaTeX, but `fig.cap` can also be used for plots in HTML, in which case the caption is written in the `<img />` tag as the `title` and `alt` attributes. Below is an example to create a figure environment in LaTeX:

```
<<waiting, fig.cap='Waiting time: Old Faithful geyser.'>>=
hist(faithful$waiting, main = "")
@
```

The LaTeX output will be:

```
\begin{figure}[]
  \includegraphics{figure/waiting}
  \caption[Waiting time]{Waiting time:
    Old Faithful geyser.} \label{fig:waiting}
\end{figure}
```

If it were a code chunk in HTML, it would have produced:

```
<img src = "figure/waiting.png"
    title = "Waiting time: Old Faithful geyser."
    alt = "Waiting time: Old Faithful geyser." />
```

## 7.8   Figure Path

We have introduced the graphical devices, but have not explained how the plots are really saved as files. Each plot is saved as a file, with the file type depending on the graphical device. The filename is determined by three chunk options: the chunk label, `fig.path`, and `fig.ext`. The `fig.path` option specifies the path of the figure (by default is a relative directory figure/), and `fig.ext` specifies the filename extension of the plot file (by default it is automatically derived from the dev option, e.g., the extension corresponding to the `Cairo_pdf` device is pdf). Strictly speaking, `fig.path` is a path prefix, e.g., `fig.path = 'figure/mcmc-'` will make all plot files have a prefix mcmc- under the figure/ directory.

All plot files in a chunk are named sequentially, from `foo-1`, `foo-2`, ..., to `foo-n`, where `foo` is the chunk label, and n is the total number of plots in the chunk. Even if a chunk only has one plot, its filename will still have the suffix `-1`.

If `fig.path` contains a directory that does not exist, **knitr** will try to create the directory automatically. For LaTeX output, only alphanumeric characters, hyphen (-), and underscore (_) are allowed in figure paths and filenames, and all other characters will be replaced by underscores. This is because LaTeX might have trouble with these characters (e.g., spaces and dots).

In most cases, we do not need to specify `fig.ext`, but when we use a custom device to save graphics, **knitr** will not be able to know the appropriate filename extension, and we have to explicitly set this option as a character string.

We emphasized the uniqueness of chunk labels in Section 5.1, and this is one reason why it has to be unique: the chunk label is used in the filenames of plots; if there are two chunks that share the same label, the latter chunk will override the plots generated in the previous chunk. The same is true for cache files in the next chapter.

# 8

## Cache

One challenge of dynamic documents is that some code chunks may take a long time to run, and these chunks may not be modified or updated frequently. In this case, caching can be very helpful. The basic idea is, a chunk will not be re-executed as long as it has not been modified since the last run, and old results will be directly loaded instead.

## 8.1 Implementation

Cache is not a new idea — both the packages **cacheSweave** and **weaver** have implemented it based on Sweave, with the former using **filehash** and the latter using *.RData images; **cacheSweave** also supports lazy-loading of objects based on **filehash**. The **knitr** package directly uses internal base R functions to save (*tools:::makeLazyLoadDB()*) and lazy-load objects (*lazyLoad()*).

The **cacheSweave** vignette has clearly explained the concept of lazy-loading. Roughly speaking, lazy-loading means an object will not be loaded into memory until it is really used anywhere — only a "promise" is created instead, which is usually fast and cheap in terms of memory consumption; when this promise is to be used for computation, the real object will be loaded from a hard disk. This is very useful for cache; sometimes we read a large object and cache it, then take a subset for analysis and this subset is also cached; in the future, the initial large object will not be loaded into R if our computation is only based on the object of its subset. For more details about promises in R, see ?promise.

To turn on caching, we can set the chunk option cache to TRUE (default is FALSE). Below is a code chunk that quickly shows the effect of cache:

```
x <- 1
Sys.sleep(10)
x <- 2
```

We used *Sys.sleep()* to let R sleep for 10 seconds. We can see the pause the first time this chunk is compiled, but when we compile it again, there will be no pause, because the code evaluation is actually completely skipped. There is an object x created in this chunk, and it will be lazy-loaded next time; **knitr** will figure out all newly created objects in a chunk and save them to lazy-load databases (*.rdb and *.rdx files). Now we can check the value of x:

```
x   # value from previous chunk

## [1] 2
```

## 8.2   Write Cache

The path of cache files is determined by the chunk option cache.path; by default all cache files are created under a directory cache/ relative to the current working directory. If the option value contains a directory (e.g., cache.path = 'cache/abc-'), cache files will be stored under that directory. Similar to figure paths, the cache directory will be automatically created if it does not exist, and cache.path can also be a prefix for cache files instead of a physical path.

The cache is invalidated and purged on any changes to the code chunk, including both the R code and chunk options; this means old cache files of this chunk are removed and replaced by new cache files. Cache filenames are identified by the chunk label as the prefix (recall that chunk labels have to be unique in a document), and the suffix of cache filenames is an MD5 hash string of an R object, which is a list including the R code, chunk options, and the value getOption('width'). The MD5 hash is calculated by the **digest** package, and it will be clear how it works by the example below, which emulates the cache filename generation in **knitr**:

```
d <- digest::digest
## imagine x$code is the code chunk; x$options are chunk
## options
x <- list(code = "1+1", options = list(results = "asis",
    fig.height = 3), width = getOption("width"))
d(x)

## [1] "667308d70fc72f26eb7454dde04af9a0"
```

```
x$code <- "1 + 1"   # add spaces to code
d(x)

## [1] "e903b616477cfa3e2314a3da65062dfb"

x$options$eval <- FALSE   # add option eval as FALSE
d(x)

## [1] "8decb2a180f7f49b47de54bd5ec8fb34"

x$width <- 40
d(x)

## [1] "7e1d77987b195b14d9b563b9a8f0ca6c"
```

The character strings of width 32 above are MD5 hashes. We can see that an MD5 hash is sensitive to changes in content. Any change will lead to a new hash string, even if the change is simply a white space. The cache filenames are of the form label_hash.rdb. Each time, **knitr** will compare the hash of the current chunk to the cache filenames; if they do not match, it means there has been a change in the chunk, and the old cache should be purged.

One exception is the include option, which is not cached because include = TRUE / FALSE does not affect code evaluation, so we can change this chunk option without affecting cache.

The reason that getOption('width') affects cache is that it may affect the width of printed text output.

## 8.3   When to Update Cache

It may not be clear when to update cache in certain circumstances, although the three components described above seem to be reasonable to take into consideration. Let's consider two cases as follows:

1. R is still being updated every few months, with each new version fixing bugs and introducing new features; should we update cache after we upgrade R to a newer version? (similar concern applies to R packages)

2. If we read an external data file in a source document, and that file has been modified; how can we tell **knitr** that all the

cached results need to be updated (even if the source document is not changed)?

In these cases, we need to put more components into the object to calculate the hash. Since a code chunk can accept arbitrary options (not only the options introduced in this book), and all chunk options are reflected in the hash, we can use additional chunk options to affect the cache.

To answer the first question, we can add a chunk option, say, `version` to the document, which takes the version of R as its value, e.g.,

```
<<cache-rversion, cache=TRUE, version=R.version.string>>=
# code which may be affected by R version
R.version.string
```

```
## [1] "R version 3.2.0 (2015-04-16)"
```

```
@
```

Then if R has been upgraded, this chunk will be re-executed.

To solve the second problem, we need to let **knitr** know changes in external files. One natural indicator is the modification time of files, which can be obtained by the function *file.info()*. Suppose the data file is named iris.csv, and we can put its modification time in a chunk option `iris_time`, e.g.,

```
<<itime, cache=TRUE, iris_time=file.info('iris.csv')$mtime>>=
# data will be re-read if iris.csv becomes newer
iris <- read.csv("iris.csv")
@
```

There are no fixed rules about when or whether to update cache; it is up to the specific applications; e.g., we do not have to purge cache after R has been upgraded. Anyway, we need to set up chunk options carefully to guarantee the results are always up-to-date.

## 8.4   Side Effects

In computer science, a side effect refers to a state change that occurs outside of a function that is not the returned value. Common side effects include creating a plot (window or file), writing a file, and printing results to the console, etc. Side effects are not straightforward to be cached — we can easily save an R object into the cache database, but it is unclear how to save a plot window because it is not a value returned by a function. For this reason, packages like **weaver** and **cacheSweave** do not cache side effects, but **knitr** will try to preserve some side effects, such as:

1. printed results: meaning that any output of a code chunk will be loaded into the output document for a cached chunk, even if it is not really evaluated. The reason is **knitr** also caches the output of a chunk as a character string. Note this means graphics output is also cached since it is part of the output;

2. loaded packages: after the evaluation of each cached chunk, the list of packages used in the current R session is written to a file under the cache path with a suffix ___packages; next time, if a cached chunk needs to be rebuilt, these packages will be loaded first. The reasons for caching package names are: it can be slow to load some packages, and a package might be loaded in a previous cached chunk that is not available to the next cached chunk when only the latter needs to be rebuilt. Note that this only applies to cached chunks, and for uncached chunks, you must always use *library()* to load packages explicitly;

3. the random seed: if a chunk created a random seed (an integer vector), the seed will be saved and loaded next time to improve reproducibility of random simulations (also see Section 12.4.7).

Although **knitr** tries to keep some side effects, there are still other types of side effects like setting *par()* or *options()* that are not cached. Users should be aware of these special cases, and make sure to clearly separate the code that is not supposed to be cached into uncached chunks, e.g., set all global options in the first chunk of a document and do not cache that chunk. Normally we have this chunk as the first chunk of a document:

```
<<setup, cache=FALSE, include=FALSE>>=
# set up some global options for the document
options(width = 60, show.signif.stars = FALSE)
# also set up global chunk options
library(knitr)
opts_chunk$set(fig.width = 5, fig.height = 4, tidy = FALSE)
@
```

In the above chunk, cache = FALSE is often unnecessary because it is the default; we can put it there if we are conservative and want to make sure this chunk is indeed not cached.

## 8.5 Chunk Dependencies

Sometimes a cached chunk may need to use objects from other cached chunks, which can bring about a serious problem — if objects in previous chunks have changed, this chunk will not be aware of the changes and will still use old cached results, unless there is a way to detect such changes from other chunks. Therefore we have to introduce dependencies into cached chunks.

### 8.5.1 Manual Dependency

There is a chunk option called dependson in **knitr** (idea taken from **cacheSweave**), which specifies which other chunks this chunk depends on by setting a vector of chunk labels like dependson = c('chunkA', 'chunkB'). Then each time either of the *cached* chunks chunkA or chunkB is rebuilt, this chunk will lose its cache and be rebuilt as well.

Chunk dependencies can form a chain; in the following example, chunkC depends on chunkB, which in turn depends on chunkA:

```
<<chunkA>>=
x <- 1
<<chunkB, dependson='chunkA'>>=
y <- x + 2
<<chunkC, dependson='chunkB'>>=
y + 5
@
```

The dependency is necessary because chunkC uses the object y that was created in chunkB, and chunkB needs the value of x created in chunkA. When x in the first chunk is changed, the latter two chunks have to be updated accordingly.

The option dependson can also take an integer vector of chunk indices, e.g., dependson = 1 means this chunk depends on the first chunk in the document, and dependson = c(3, 5) indicates dependency on the third and fifth chunks. If the indices are negative, it means counting backwards from this chunk. For example, dependson = -1 means this chunk depends on the previous chunk, and -c(1, 2, 3) means the previous three chunks. Note that when dependson takes integer values, it cannot make a chunk depend on later chunks (only previous chunks are possible candidates); character values of dependson do not have this restriction.

### 8.5.2 Automatic Dependency

Another way to specify the dependencies among chunks is to use the chunk option autodep and the function *dep_auto()*. This is an experimental feature borrowed from **weaver**, which frees us from setting chunk dependencies manually. The basic idea is, if a latter chunk uses any objects created from a previous chunk, the latter chunk is said to depend on the previous one.

The function *findGlobals()* in the **codetools** package is used to find out all global objects in a chunk, and according to its documentation, the result is an approximation. Global objects roughly mean the ones that are not created locally, e.g., in the expression function() {y <- x}, x must be an existing global object outside (no matter what object it really is) because we do not see its creation in the body of this function, whereas y is local. Meanwhile, we also need to save the list of objects created in each cached chunk, so that we can compare them to the global objects in latter chunks. For example, if chunk A created an object x and chunk B uses this object, chunk B must depend on A, i.e., whenever A changes, B must also be updated.

When autodep = TRUE, **knitr** will write out the names of objects created in a cached chunk as well as those global objects in two files named ___objects and ___globals, respectively; later we can use the function *dep_auto()* to analyze the object names to figure out the dependencies automatically. A typical use is:

```
<<setup, cache=FALSE, include=FALSE>>=
opts_chunk$set(autodep = TRUE)  # set autodep globally
dep_auto()  # figure out dependencies
@
```

Yet another way to specify dependencies is *dep_prev()*: this is a conservative approach that sets the dependencies so that a cached chunk will depend on all its previous chunks, i.e., whenever a previous chunk is updated, all later chunks will be updated accordingly.

In any case, dependency on *uncached* chunks is meaningless to **knitr**, because **knitr** only checks changes for cached chunks; **knitr** will give a warning when it sees dependency on uncached chunks. If we have to depend on uncached chunks at all, we can use the trick introduced in Section 8.3, i.e., to put the uncached objects in the chunk options of cached chunks. Below is an example:

```
<<A, cache=FALSE>>=
x <- 1
@


<<B, cache=TRUE, foo=x>>=
y <- x + 2
@
```

We created an object x in an uncached chunk A, and used it in a cached chunk B. If there is no dependency between the two chunks, B will not update when A is updated, but if we have set an option foo = x in chunk B, B will automatically be updated if the value of x has changed, which leads to changes in B's chunk options.

## 8.6   Load Cache Manually

Usually the cache database is automatically loaded for a cached chunk, and we can actually load it manually. This has a useful application: imagine you calculated a value x in a later chunk, but you want to use it earlier in the document. That is not possible because **knitr** compiles the document in a linear fashion, and you cannot use an object created in the future. However, if you have turned on the cache for that chunk, you may just load its cache database early.

The function *load_cache()* in **knitr** was designed for this purpose. It takes a chunk label to find the cache database, and optionally you can specify the object that you want this function to return from the cache.

```
load_cache(label, object, notfound = "NOT AVAILABLE",
    path = opts_chunk$get("cache.path"), lazy = TRUE)
```

Now suppose you have a cached chunk named foo later in the document, which creates an object x, you can load_cache('foo', 'x') to fetch the value of x in that chunk. Of course, the first time you compile the document, x will not be available, and that is what the argument notfound is for. If you use x in an inline R expression, you will see NOT AVAILABLE in the output, and it will be replaced by the value of x after you compile the document again, since the chunk foo has been cached.

## 8.7 Other Options

Although lazy-loading is useful, it may not work in certain cases for reasons that are still not clear to us. Anyway, you can turn off lazy-loading using the chunk option cache.lazy = FALSE. In this case, **knitr** will just save the objects with *save()*, and load them with *load()*, which should always work.

Sometimes you may be tweaking comments in code without really changing other parts of the code, and you certainly do not want to update the cache database just because you updated the code comments. In this case, you can use the chunk option cache.comments = FALSE. Then comments will be excluded when calculating the MD5 hash, and therefore changes in comments will not affect the cache.

# 9

# *Cross Reference*

We can cross reference both code chunks and child documents in **knitr**. This enables us to better organize our source documents. Below is a practical example: we have a custom **ggplot2** theme and we want to apply it to a few plots in the document.

```
<<my-theme, eval=FALSE>>=
theme(legend.text = element_text(size = 12, angle = 45)) +
    theme(legend.position = "bottom")
@
```

If we were to use this piece of code only once, we can just copy and paste it to the code chunk, but it is certainly not a good idea to paste it to multiple chunks, since it will be a disaster to maintain. We can simply use a reference to it using its chunk label, e.g.,

```
qplot(carat, price, data = diamonds, color = cut) +
    <<my-theme>>
```

Then **knitr** will expand <<my-theme>> to the real source code before evaluating this chunk. We can use this reference in multiple places but only maintain one copy of the source.

## 9.1 Chunk Reference

With chunk references, we can easily reuse code chunks without typing them again. We can embed a defined chunk into another chunk, or just reuse a whole chunk as a new chunk.

### 9.1.1 Embed Code Chunks

One chunk can be used as a part of another chunk, and the syntax is <<label>> (white spaces are allowed before it; label means the chunk label); note there is no = after >> like chunk headers. For example, we embed chunk A in B:

```
<<A>>=
x <- rnorm(1)
@
<<B>>=
x
<<A>>
x
@
```

In this case, chunk B is essentially this (<<A>> is replaced by the code in chunk A but note all chunk options in A are ignored, including `eval`):

```
x
x <- rnorm(1)
x
```

Chunks can be nested recursively within each other as long as the recursion is finite, e.g., we embed A into B, and B into C, but we must not embed C into A again, otherwise there will be infinite recursion.

### 9.1.2   Reuse Whole Chunks

There are two ways to reuse a whole chunk. The first one is to use the same label but leave the chunk empty. One problem with this approach is that we cannot cache both chunks if their chunk options are different because their MD5 hashes will be different, and **knitr** only allows one set of cache files per label. Here is one example:

```
<<chunkA, eval=FALSE>>=
x <- 1 + 1
@
<<chunkA, eval=TRUE>>=
@
```

The second approach is to use the `ref.label` option, which takes a vector of the chunk labels of source chunks. We can use a new label for the target chunk. In the following example, chunk C uses code from both A and B:

```
<<A>>=
x <- rnorm(1)
@
<<B>>=
y <- x + 2
@
<<C, ref.label=c('A', 'B')>>=
@
```

The code for chunk C is essentially this:

```
x <- rnorm(1)
y <- x + 2
```

## 9.2 Code Externalization

It can be more convenient to write R code chunks in a separate R script, rather than mixing them into a source document; for example, we can run R code successively in a pure R script from one chunk to the other without jumping through other text.

The other reason is that some editors such as LyX do not have support to run R code interactively, and we have to recompile the whole document each time, even if we only want to know the results of a single chunk.

Therefore **knitr** introduced the feature of code externalization: code chunks can be read from an external R script via *read_chunk()*. The R script can be written in two forms: we either use labels in the script to separate code chunks, or specify chunks based on line numbers.

### 9.2.1 Labeled Chunks

The setting is like this: the R script also uses chunk labels (marked in the form ## ---- chunk-label); if the code chunk in the source document is empty, **knitr** will match its label with the label in the R script to input external R code.

For example, suppose this is a code chunk labelled as Q1 in an R script named shared.R, which is under the same directory as the source document:

```
## ---- Q1 ----
gcd <- function(m, n) {
    while ((r <- m%%n) != 0) {
        m <- n
        n <- r
    }
    n
}
```

In the source document, we can first read the script using the function *read_chunk()*:

```
read_chunk("shared.R")
```

This is usually done in an early chunk such as the first chunk of a document, and we can use the chunk Q1 later in the source document:

```
<<Q1>>=
@
```

## 9.2.2   Line-Based Chunks

By default, *read_chunk()* assumes that the R script is labeled (## ----
is the delimiter), and there is an alternative approach to specify code chunks via the three arguments labels, from, and to, which are vectors of the same length. The starting and ending line numbers of code chunks can be set through from and to, respectively, and labels is a vector of chunk labels.

For example, if we want the lines 1-5, 7-9, and 15-21 in the R script foo.R to form three chunks with labels A, B, and C, we can call the function *read_chunk()* like this:

```
read_chunk("foo.R", labels = c("A", "B", "C"), from = c(1,
    7, 15), to = c(5, 9, 21))
```

Then we can write three empty chunks in the source document, with labels A, B, and C. Alternatively, from and to can be regular expressions for the starting and ending lines.

Different documents can read the same R script, so the R code can be reusable across different input documents.

## 9.3 Child Documents

The concept of child documents should be familiar to LATEX users —
when the main document is large, we can split it into smaller parts and
input them into the main document using \input{foo.tex}. For example, a book can be split into chapters, with each chapter in one file.

### 9.3.1 Input Child Documents

Similarly, we can manage a **knitr** source document as a collection of
child documents. The chunk option child provides a reference to child
documents. Suppose we have a main document named book.Rnw, and
a child document named chap1.Rnw under the same directory. In the
main document, we have:

```
Here is one chunk in the main document.

<<A, eval=TRUE>>=
x <- rnorm(12)
@

We include a child document which uses the variable x.

<<B, child='chapt1.Rnw'>>=
@

One realization of a Chi-square random variable
with df 12 is \Sexpr{y}.
```

We referenced the child document in chunk B. When the main document is compiled, **knitr** will look for the child document and compile
it accordingly; everything in the environment of the main document up
to this point will be available to the child document, e.g., the variable x.
The child document is:

```
This is a child document.

<<B1>>=
y <- sum(x^2)
@
```

We created a new object y in the child document; after the child
document has been compiled, it will be available to the later chunks in

the main document as well. That is why \Sexpr{y} will work. As a side
note, the sum of $n$ i.i.d standard Normal random variables follows the
$\chi^2_n$ distribution (with $n$ degrees of freedom), so y is one random number
generated from $\chi^2_{12}$.

Like chunk references, child documents have no limits on the levels
of nesting. One child document can have further children documents,
and one chunk can include more than one child document.

### 9.3.2  Child Documents as Templates

It is common to do the same analysis using a template with different
data input, and child documents can be helpful for such tasks as well.
As a trivial example, we continue to generate another random number
from the Chi-square distribution in the main document:

```
% second part of book.Rnw
Continue the above example. Now we change the degrees
of freedom to 8.

<<C, eval=TRUE>>=
x <- rnorm(8)
@

And include the child document again.

<<D, child='chapt1.Rnw'>>=
@

One realization of a Chi-square random variable
with df 8 is \Sexpr{y}.
```

What the child document does here is only to calculate the sum of
squares for x and assign the result to y. It is very similar to a sub-
routine, even though it is not "pure source code" as we usually see.

With chunk references and child documents, we can modularize an
analysis in the same manner of programming.

### 9.3.3  Standalone Mode

This section is specific to LaTeX. Rnw child documents are often incomplete in the sense that they do not have the LaTeX preamble (lines from
\documentclass to \begin{document}), so if we compile them directly,
we will end up with LaTeX errors.

Although child documents are supposed to be related to the parent document, it is not necessarily true in some cases. Sometimes a child document is there only for the purpose of organizing a huge document, and the computation in the child document may be completely irrelevant to the parent. In this case, all we need is to borrow the preamble of the parent document and append it to the child document when compiling the results.

The function *set_parent()* notifies **knitr** of the parent document of a child; once this function is called, **knitr** will read the preamble of the parent document and write it to the child document when an Rnw document is compiled to TEX. For example, we can do this in chapt1.Rnw:

```
<<parent, include=FALSE>>=
set_parent("book.Rnw")
@
```

Then, whatever LATEX styles are defined in the preamble of book.Rnw will be available to chapt1.tex as if the content of chapt1.Rnw were in book.Rnw.

# 10

---

## *Hooks*

---

Hooks are an important component to extend **knitr**. A hook is a user-defined R function to fulfill tasks beyond the default capability of **knitr**. There are two types of hooks: *chunk* hooks and *output* hooks. We have already introduced some built-in output hooks in Section 5.3, and how to customize both the chunk and inline R output. In this chapter we focus on chunk hooks.

---

## 10.1   Chunk Hooks

A chunk hook is a function stored in `knit_hooks` and triggered by a custom chunk option. All chunk hooks have three arguments: `before`, `options`, and `envir` (explained later).

### 10.1.1   Create Chunk Hooks

A chunk hook can be arbitrarily named, as long as it does not clash with existing hooks in `knit_hooks`. Names of all built-in hooks are:

```
names(knit_hooks$get(default = TRUE))

## [1] "source"   "output"    "warning"  "message"
## [5] "error"    "plot"      "inline"   "chunk"
## [9] "text"     "document"
```

For example, the name `margin` is not in the above names, so we can name a chunk hook as `margin`:

```
knit_hooks$set(margin = function(before, options, envir) {
    if (before)
        par(mar = c(4, 4, 0.1, 0.1)) else NULL
})
```

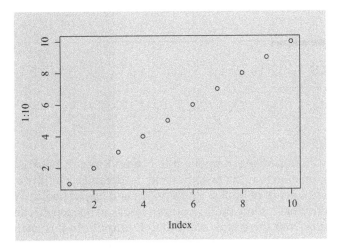

FIGURE 10.1: A plot with the default margin, i.e., `par(mar = c(5.1, 4.1, 4.1, 2.1))`.

This hook is used to set the margin parameter with *par()* for R base graphics (because the default margin is often too big).

### 10.1.2 Trigger Chunk Hooks

After we have defined a hook, we need to set a chunk option with the same name to a non-NULL value in order to execute the hook function. By default all undefined chunk options are NULL, so the chunk below is equivalent to a chunk with the option margin = NULL, which will not call the hook we just defined when the chunk is compiled (Figure 10.1):

```
<<mar-normal>>=
par(bg = "gray")
plot(1:10)
@
```

However, when we set margin = TRUE, the hook will be called before the chunk is evaluated because TRUE is not NULL (Figure 10.2):

```
<<mar-small, margin=TRUE>>=
par(bg = "gray")
plot(1:10)
@
```

We set the plot background to be gray just to show the margins more clearly.

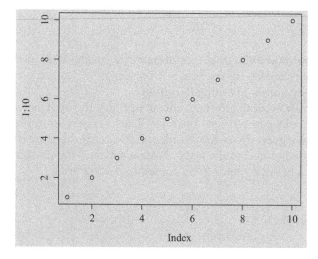

FIGURE 10.2: A plot with a smaller margin using the `margin` hook
(`par(mar = c(4, 4, .1, .1))`).

### 10.1.3 Hook Arguments

Now we explain the four arguments of a chunk hook. Note all four
arguments are optional.

**before** a logical value: `TRUE` if the hook is called before a chunk, and
`FALSE` when a hook is called after a chunk

**options** a list of current chunk options, e.g., `options$label` is the current chunk label

**envir** the environment in which the current code chunk is evaluated,
e.g., `envir$x` is the object x in the current chunk (if it exists)

**name** the name of the current hook function

A chunk is called twice for a chunk: once before a chunk and once after
a chunk. In the above `margin` hook, *par()* was called before a chunk is
evaluated, so the plots will use the parameters set by *par()*. If we set
*par()* after a chunk, it will be too late (hence useless) because the plots
have already been drawn.

### 10.1.4 Hooks and Chunk Options

Since chunk hooks are called as long as the corresponding chunk options are not `NULL`, we can set these chunk options globally if we want
the chunk hooks to be applied to all chunks in a document, e.g.,

```
opts_chunk$set(margin = TRUE)
```

Note that non-NULL does not necessarily mean TRUE; in the above example, we can also set margin = 1 or margin = 'hello', and so on, because these values are not NULL either.

Since **knitr** accepts arbitrary chunk options, the options argument in chunk hooks can be very flexible. The previous example did not actually make good use of the chunk option margin, because this option was basically ignored in the hook. Now we extend the hook a little bit, with margin being a vector to be passed to *par(mar = ...)*:

```
knit_hooks$set(margin = function(before, options, envir) {
    if (before) {
        m <- options$margin
        if (is.numeric(m) && length(m) == 4L) {
            par(mar = m)
        }
    } else NULL
})
```

Instead of using a fixed value c(4, 4, .1, .1) for the margin parameter, we can use any numeric vectors of length 4 now, e.g.,

```
<<mar-numeric, margin=c(2, 3, 1, .1)>>=
plot(1:10)
@
```

Then before this chunk is evaluated, par(mar = c(2, 3, 1, .1)) will be called first.

### 10.1.5  Write Output

Since a chunk hook is a function, it also has a returned value. If the value returned is character, it will be written to the output. The previous hooks did not write anything to the output because they did not return character values (*par()* returns a list).

Below is a hook that returns character values: a down brace before a chunk and an up brace ⌣ after a chunk.

```
knit_hooks$set(brace = function(before, options, envir) {
    if (before) {
        "\\noindent\\downbracefill{}\n\n"
```

```
    } else {
        "\n\n\\noindent\\upbracefill{}\n"
    }
})
```

We apply this `brace` hook to the following chunk:

```
<<test, brace=TRUE>>=
1 + 1

## [1] 2

rnorm(10)

##   [1] -0.1738  1.1675  0.8677 -0.8149 -1.6213  0.8553
##   [7] -1.8358 -0.7550 -1.6286 -0.6447

@
```

Chunk hooks that return character values allow us to write anything we want to the chunk output. One important application is to write images to the output, which we have created through R code in the chunk. The character values may be like \includegraphics{...} (LATEX), <img src='...' /> (HTML) or ! [] (...) (Markdown), etc. This is the trick we will use for the next few sections, such as saving **rgl** and GGobi plots.

## 10.2 Examples

In this section we give some examples of chunk hooks, most of which have been predefined in **knitr**, i.e., we can use them directly after **knitr** has been loaded.

### 10.2.1 Crop Plots

Some R users may have been suffering from the extra white margins in R plots, especially in base graphics (**ggplot2** is usually better in this aspect). The default graphical option `mar` is about `c(5, 4, 4, 2)` as

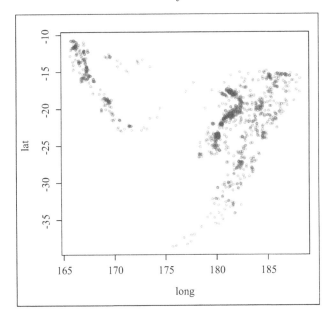

FIGURE 10.3: The original plot produced in R, with a large white margin.

we mentioned in Figure 10.1 (also see ?par), which is often too big. Instead of endlessly tweaking par(mar), we may consider the program pdfcrop, which can crop the white margin automatically (http://www. ctan.org/pkg/pdfcrop). In **knitr**, we can set up the hook *hook_pdfcrop()* to work with a chunk option, say, crop.

```
knit_hooks$set(crop = hook_pdfcrop)
```

Now, we compare two plots produced by the same code chunk below. The first one is not cropped (Figure 10.3); then the same plot is produced but with a chunk option crop = TRUE, which will call the cropping hook (Figure 10.4).

```
par(mar = c(5, 4, 4, 2))  # large margin
plot(lat ~ long, data = quakes, pch = 20, col = rgb(0, 0,
    0, 0.2))
```

As we can see, the white margins are gone (to better see the difference, we have put a frame box around each plot). If we use *par()*, it might be hard and tedious to figure out a reasonable amount of margin

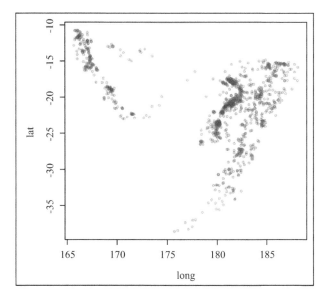

FIGURE 10.4: The cropped plot; obviously the white margins on the top and right have been removed.

such that no label is cropped due to a too-small margin, nor do we get too large a margin.

### 10.2.2 rgl Plots

With the hook *hook_rgl()*, we can easily save snapshots from the **rgl** package (Adler and Murdoch, 2014). The rgl hook is a good example of taking care of details by carefully using the options argument in the hook; for example, we cannot directly set the width and height of rgl plots in *rgl.snapshot()* or *rgl.postscript()*, so we make use of the options fig.width, fig.height, and dpi to calculate the expected size of the window, then resize the current window by *par3d()*, save the plot, and finally return a character string containing the appropriate code to insert the plot into the output. Here is a quick and dirty version of *hook_rgl()*:

```
knit_hooks$set(rgl = function(before, options, envir) {
    library(rgl)
    if (before || rgl.cur() == 0)
        return()  # return nothing before a chunk
    name <- paste(options$fig.path, options$label, sep = "")
```

FIGURE 10.5: An **rgl** plot captured by *hook_rgl()*: this hook function calls *rgl.snapshot()* in **rgl** to save the snapshot into a PNG image.

```
    rgl.snapshot(paste(name, ".png", sep = ""), fmt = "png")
    paste("\\includegraphics{", name, "}\n", sep = "")
})
```

The real hook function in **knitr** is much more complicated than this due to a lot of details to be taken into consideration. Below is an example of how to save **rgl** plots using the rgl hook. First we define a hook named rgl for the function *hook_rgl()*:

```
knit_hooks$set(rgl = hook_rgl)
```

Then we only have to set the chunk option rgl = TRUE and the captured plot is shown in Figure 10.5.

```
library(rgl)
demo("bivar", package = "rgl", echo = FALSE)
par3d(zoom = 0.7)
```

### 10.2.3   Manually Save Plots

We have explained how R plots are recorded in Section 7.2. In some cases, it is not possible to capture plots by *recordPlot()* (such as **rgl** plots), but we can save them using other functions. To insert these plots into the output, we need to set up a hook first like this (see the help page ?hook_plot_custom for details):

FIGURE 10.6: A plot created and exported by GGobi, and written into LaTeX by the hook *hook_plot_custom()*.

```
knit_hooks$set(custom_plot = hook_plot_custom)
```

Then we set the chunk option `custom_plot = TRUE`, and manually write plot files in the chunk. Here we show an example of capturing GGobi plots using the function *ggobi_display_save_picture()* in the **rggobi** package (Temple Lang et al., 2014):

```
<<ggobi-plot, custom_plot=TRUE, fig.ext='png'>>=
library(rggobi)
data("flea", package = "tourr")
ggobi(flea)
Sys.sleep(1)  # wait for snapshot
ggobi_display_save_picture(path = fig_path(".png"))
@
```

Figure 10.6 is the plot output from GGobi. Two things to note here are:

1. we have to make sure the plot filename is from *fig_path()*, which is a convenience function to return the figure path for the current chunk (a combination of the chunk label, `fig.path` and `fig.ext`);

2. we need to set the chunk option `fig.ext` (figure file extension) because **knitr** will be unable to figure out its value automatically (we are not using any graphical devices).

We can even save a series of images to make an animation with the
option `fig.show = 'animate'` (Section 7.3.1); below is an example of
zooming into a scatterplot using **rgl** (for the real animation, see **knitr**'s
main manual):

```
## use chunk options: custom_plot=TRUE, fig.ext='png',
## out.width='2.5in', fig.show='animate', fig.num=20
library(animation)   # adapted from demo('rgl_animation')
data(pollen)
uM <- matrix(c(-0.37, -0.51, -0.77, 0, -0.73, 0.67, -0.1,
    0, 0.57, 0.53, -0.63, 0, 0, 0, 0, 1), 4, 4)
library(rgl)
open3d(userMatrix = uM, windowRect = c(0, 0, 400, 400))
plot3d(pollen[, 1:3])
zm <- seq(1, 0.05, length = 20)
par3d(zoom = 1)   # change the zoom factor gradually later
for (i in 1:length(zm)) {
    par3d(zoom = zm[i])
    Sys.sleep(0.05)
    rgl.snapshot(paste(fig_path(i), "png", sep = "."))
}
```

### 10.2.4   Optimize PNG Plots

The free software OptiPNG is a PNG optimizer that re-compresses image files to a smaller size, without losing any information (`http://
optipng.sourceforge.net/`). In **knitr**, the hook function *hook_optipng()*
is a wrapper around OptiPNG to compress PNG plots, and OptiPNG
has to be installed beforehand; for Windows users, the executable has
to be in the `PATH` variable. We can set up the hook as usual:

```
knit_hooks$set(optipng = hook_optipng)
```

Then we can either set the chunk option `optipng = TRUE` to enable
it for a chunk, or pass a character string to this option so that it is used
by OptiPNG as additional command line arguments. For example, we
can use `optipng = '-o7'` to specify the highest level of optimization.
See the documentation of OptiPNG for all possible arguments.

FIGURE 10.7: Adding elements to an existing rgl plot: if we do not open a new device, latter elements will be added to the existing device.

### 10.2.5 Close an rgl Device

The default rgl hook *hook_rgl()* does not close the rgl device before drawing a new plot, which may be problematic, because the latter plot is drawn on the previous scene. For example, we get one plot with two spheres (Figure 10.7) when we execute the following two lines together, but two plots with one sphere in each if we close the first plot and run the second line:

```
rgl.spheres(0, 0, 0)
rgl.spheres(0, 2, 0)
```

Normally different code chunks use different graphical devices, so graphical elements in a latter chunk will not be added to a previous chunk, but this is not true for rgl plots. In order to close the device before drawing plots, we have to tweak the hook a little bit, e.g.,

```
knit_hooks$set(rgl = function(before, options, envir) {
    # if a device was opened before this chunk, close it
    if (before && rgl.cur() > 0)
        rgl.close()
    hook_rgl(before, options, envir)
})
```

The function *rgl.cur()* returns the current device id; if it is greater than 0, it means there is an existing device, and we can close it by *rgl.close()*.

**10.2.6    WebGL**

We introduced how to save static **rgl** plots in Section 10.2.2. In fact, we can also export the **rgl** 3D plot into WebGL (`http://en.wikipedia.`  `org/wiki/WebGL`) using the *writeWebGL()* function, so that the plot can be reproduced in a Web browser that supports WebGL. For example, we can rotate and zoom in/out the plot.

   The hook function *hook_webgl()* in **knitr** is a wrapper to the WebGL function in **rgl**. With this hook, we can capture a 3D scene into the HTML output.

# 11

## Language Engines

We can work with a lot of languages and tools in **knitr**, including but not limited to R, although **knitr** is an R package and has to be run within the R environment in the first place. Currently **knitr** supports Python, Ruby, Haskell, awk/gawk, sed, shell scripts, Perl, SAS, TikZ, Graphviz, and C++, etc. We have to install the corresponding software package in advance to use an engine.

## 11.1 Design

Like chunk hooks, all language engines are essentially R functions in **knitr**. These functions pass the code chunk to external programs, run the code there, get the results back, and write to the output. In most cases, the code is passed to external programs via the *system()* function. For example, we can pass code to bash via the -c option.

```
system("bash -c 'ls ~ | grep ^D'", intern = TRUE)

## [1] "Desktop"   "Downloads" "Dropbox"
```

For those who are not familiar with bash scripts, the code ls ~ | grep ^D means to list files under the home directory (~) and pass the filenames to grep through the pipe (|) to match those starting with the letter D; ls and grep are standard Linux commands.

The chunk option engine can be used to specify the language engine for a chunk, e.g., the chunk below uses engine = 'bash':

```
ls ~ | grep ^D

## Desktop
## Downloads
## Dropbox
```

111

Then the code in the chunk will be treated as a bash script instead of an R script. The output rendering is similar to R output: the source code is passed to the `source` hook (i.e., `knit_hooks$get('source')`), and the output is passed to the `output` hook (`knit_hooks$get('output')`). The built-in output hooks are fairly general in terms of document formats; we do not need to think about whether the output is to be LaTeX or HTML or Markdown; everything will be automatically and properly marked up according to the output document format.

### 11.1.1   The Engine Function

All language engines are stored in the object `knit_engines`, which has the `$get()` and `$set()` methods like `knit_hooks` (chunk hooks) and `opts_chunk` (chunk options); e.g., we can get the Python engine by `knit_engines$get('python')`, or override the built-in Python engine by `knit_engines$set(python = function(options) {...})`.

An engine has one argument: `options`, which is a list of current chunk options. Among all options there is one special option named `code`, which is the code (as a character string) of the current chunk and plays the central role in the language engine.

To continue the bash example, we can define a preliminary engine like this:

```
knit_engines$set(bash = function(options) {
    code <- paste(options$code, collapse = "\n")
    out <- system(paste("bash -c", shQuote(code), sep = " "),
        intern = TRUE)
    paste(c(code, out), collapse = "\n")
})
```

What this engine does is to concatenate the command bash -c with the source code, execute the whole command via *system()*, and return both the source code and output as one character string separated by line breaks. The returned character string will be written into the output document.

The real bash engine is more complicated than this: it has to take care of some chunk options such as `echo`, `results`, `include`, `cache`, and so on. For example, when `echo = FALSE`, the source code should be hidden, and when `cache = TRUE`, the code chunk should be cached. In all, the behavior of these language engines is very similar to the R engine, although the support is not as comprehensive as R.

Note in particular the cache of language engines other than R: in most cases, only the side effects such as printing are cached, due to the fact that it is difficult for R to know which objects are created in a code chunk if the code is not written in R. In other words, objects are lost when we exit from a chunk (unless they are exported to files). Normally we will not be able to reuse an object created from previous chunks. The reason that we can use R objects across different chunks is that all R chunks are evaluated in the same R session, but other languages are evaluated in separate sessions per chunk basis.

### 11.1.2 Engine Options

For language engines, there are two common chunk options:

**engine.path** specifies the full path to the engine program as a character string; this may be useful to Windows users when the program to be called is not in the environmental variable PATH (i.e., the program cannot be run without full path in the command line), or to Linux users when there are multiple versions of one program installed and we do not want to use the default version; in both cases, we can set the chunk option engine.path = 'full/path/to/program', e.g., engine.path = '/usr/bin/ruby1.9.1' (if there are multiple versions of Ruby) or engine.path = 'C:/Program Files/SASHome/x86/9.3/sas.exe' (to specify the full path of SAS);

**engine.opts** additional options to be passed to an engine; its value depends on the specific engine; for most engines, it contains additional command line arguments, e.g., for engine = 'ruby', we can set engine.opts = '-v' for Ruby to print its version number, then turn on the verbose mode.

## 11.2 Languages and Tools

Most languages and tools are supported through the *system()* interface, as mentioned in the last section. There are a few exceptions, however, such as C++ and TikZ.

### 11.2.1 C++

C++ is supported in **knitr** through the **Rcpp** package (Eddelbuettel et al., 2015). When we set engine = 'Rcpp', the function *sourceCpp()* in **Rcpp** is used to compile C++ code chunks, which in fact calls R CMD

SHLIB internally to build a shared library and load it into R for future use.

Below is an example for the Fibonacci series ($x_i = x_{i-1} + x_{i-2}$, $x_0 = 0$ and $x_1 = 1$) in C++ with **Rcpp**:

```
#include <Rcpp.h>

// [[Rcpp::export]]
int fibCpp(const int x) {
    if (x == 0 || x == 1) return(x);
    return (fibCpp(x - 1)) + fibCpp(x - 2);
}
```

After it is compiled, we can call the function *fibCpp()* in R directly because we have marked it with the Rcpp::export attribute.

```
fibCpp(10L)
```

```
## [1] 55
```

```
system.time(fibCpp(27L))
```

```
##    user  system elapsed
##   0.001   0.000   0.001
```

Below is the version implemented in pure R:

```
fibR <- function(x) {
    if (x == 0L || x == 1L)
        return(x)
    return(fibR(x - 1L) + fibR(x - 2L))
}
```

Unsurprisingly, the R version is much slower, although the numeric results are the same:

```
fibR(10L)
```

```
## [1] 55
```

```
system.time(fibR(27L))
```

```
##    user  system elapsed
##   0.708   0.000   0.708
```

Finally, we can pass additional arguments to *sourceCpp()* via the chunk option engine.opts. For example, we can specify engine.opts = list(showOutput = TRUE) to show the output of R CMD SHLIB (note showOutput is an argument of *sourceCpp()*).

## 11.2.2 C/Fortran

There are two simple language engines c and fortran for the C language and Fortran, respectively. These engines are nothing but wrappers for the command R CMD SHLIB and the R function *dyn.load()*. What they do is to write the code chunk to a temporary file, run R CMD SHLIB to compile it, and use *dyn.load()* to load the compiled library (a .dll or .so file). To use these engines, you have to make sure you have the C/Fortran compilers in your system, such as GCC.

```
# the compilers in the environment in which this book
# was written
Sys.which("gcc")

##            gcc
## "/usr/bin/gcc"

Sys.which("gfortran")

##              gfortran
## "/usr/bin/gfortran"
```

Below are two examples demonstrating the usage of these two engines. First, we set the chunk option engine = 'c' for this example:

```
/* calculate the square of a number */
void my_square(double *x) {
  *x = *x * *x;
}
```

After compiling the above code chunk, we can call the C function *my_square()* via the .C() interface:

```
.C("my_square", 9)

## [[1]]
## [1] 81

.C("my_square", 123)
```

```
## [[1]]
## [1] 15129
```

Next, we show a Fortran example by setting the chunk option `engine`
= 'fortran' for the chunk below:

```
C Fortran test
      subroutine fexp(n, x)
      double precision x
C  output
      integer n, i
C  input value
      do 10 i = 1, n
         x = dexp(dcos(dsin(dble(float(i)))))
  10  continue
      return
      end
```

And we can call the Fortran sub-routine via the *.Fortran()* interface:

```
res <- .Fortran("fexp", n = 100000L, x = 0)
str(res)

## List of 2
##  $ n: int 100000
##  $ x: num 2.72
```

### 11.2.3  Interpreted Languages

C++, C, and Fortran belong to compiled languages, and there are other
languages that are interpreted languages. For these languages, we can
execute the code without compiling it. Examples include awk and shell
scripts. There are also some languages that belong to both categories,
such as Python. Table 11.1 lists some interpreted languages supported
by **knitr** via the *system()* interface.

For example, a Perl chunk is executed with `perl -e code` where
code is the character string of the code chunk. For awk and sed, the
argument after the program is treated as the source code, so they do
not need an argument name for the code, e.g., `awk 'END{print NR;}'`
`README` counts the number of lines in the file README. For SAS, the
code chunk is written into a file tempfile.sas, and executed as `sas -SYSIN`
`tempfile.sas`. There are three shell variants: `sh`, `bash`, and `zsh`.

TABLE 11.1: Interpreted languages supported by **knitr**: the language name, engine name, and the command line argument to execute code.

| Language | Engine | Code argument |
|---|---|---|
| Python | python | -c |
| Ruby | ruby | -e |
| (g)awk | (g)awk | |
| sed | sed | |
| shell | sh/bash/zsh | -c |
| Perl | perl | -e |
| Haskell | haskell | -e |
| CoffeeScript | coffee | -e |
| Groovy | groovy | -e |
| Node.js | node | -e |
| Scala | scala | -e |
| SAS | sas | -SYSIN |

As we mentioned before, the engine name itself may not be the executable, so we may need to specify the path to the real path of the program. For Haskell, haskell is not the program to run Haskell, whereas ghc is, so we need to specify both engine = 'haskell' and engine.path = 'ghc'.

We give a few examples of the above languages. Here is a Python chunk (chunk option engine = 'python'):

```
x = 'hello, python world!'
print x
print x.split(' ')

## hello, python world!
## ['hello,', 'python', 'world!']
```

Here is a Ruby chunk:

```
x = 'hello, ruby world!'
p x.split(' ')

## ["hello,", "ruby", "world!"]
```

Below is an awk script to count the number of non-empty lines in the NEWS.Rd file of the **knitr** package: in awk, NF denotes the number of fields on a line; when it is not 0, the variable i increases by 1, and that is why the script counts the non-empty lines in the file. Note that we used `engine.opts = shQuote(system.file('NEWS.Rd', package = 'knitr'))` for this chunk; i.e., we get the path to the NEWS.Rd file from R, quote it by *shQuote()*, and pass it to awk as the second argument (remember the first argument is the code chunk), which means the file to be read into awk.

```
# how many non-empty lines in the NEWS file?
NF {
  i = i + 1
}
END { print i }

## 8
```

Finally we have a Perl code chunk:

```
$test = "jello world";
$test =~ s/j/h/;
print $test

## hello world
```

### 11.2.4  Stan

We can use the **rstan** package (Guo et al., 2014) to compile models of Stan, a relatively new programming language featuring Bayesian statistical inference. There is a language engine called `stan` in **knitr** that allows us to write Stan models in code chunks. We can certainly compile a Stan model in a normal R code chunk without using a special language engine, by saving the model as a file, or writing the model as a long character string in R code. Both ways have their disadvantages: it is not convenient for the reader to see the real model in the report if it is in an external file, and it is cumbersome to write a model as a long character string of multiple lines in R. The `stan` engine makes it possible to write the model as a code chunk, which solves both problems mentioned before. Here is a simple example of sampling from the posterior distribution of the parameter $p$ (probability of $X = 1$) of a Bernoulli distribution:

```
<<engine='stan', engine.opts = list(x = 'ex1')>>=

data {
  int<lower=0,upper=1> X[20];
```

```
}
parameters {
  real<lower=0,upper=1> p;
}
model {
  X ~ bernoulli(p);
}

@
```

Besides the chunk option `engine = 'stan'`, we also specified the option `engine.opts = list(x = 'ex1')`. Here x means the name of the Stan model to be saved in the R session. This code chunk will pass the model to the function *stan_model()* in **rstan**, and save the model to the object ex1. That is why we can use the object ex1 in the next chunk:

```
library(rstan)
fit <- sampling(ex1, data = list(X = rbinom(20, 1, 0.3)))


SAMPLING FOR MODEL 'anon_model' NOW (CHAIN 1).

Iteration:    1 / 2000 [  0%]  (Warmup)
Iteration:  200 / 2000 [ 10%]  (Warmup)
Iteration:  400 / 2000 [ 20%]  (Warmup)
Iteration:  600 / 2000 [ 30%]  (Warmup)
Iteration:  800 / 2000 [ 40%]  (Warmup)
Iteration: 1000 / 2000 [ 50%]  (Warmup)
Iteration: 1001 / 2000 [ 50%]  (Sampling)
. . . .


print(fit)

Inference for Stan model: anon_model.
4 chains, each with iter=2000; warmup=1000; thin=1;
post-warmup draws per chain=1000,
total post-warmup draws=4000.

       mean se_mean   sd    2.5%     25%     50%     75%
p      0.36    0.00 0.10    0.18    0.29    0.36    0.43
lp__ -14.93    0.02 0.73  -16.99  -15.12  -14.65  -14.47
      97.5% n_eff Rhat
p      0.57  1498    1
lp__ -14.42  1703    1
. . . .
```

We generated 20 random data points from the Bernoulli distribution with $p = 0.3$, and used them as the sample data Y for the Bayesian inference. You can see from the sampling output that the posterior mean of $p$ is near 0.3.

### 11.2.5   TikZ

We introduced the **tikzDevice** package in Section 7.6, which enables us to convert R graphics to TikZ (Tantau, 2008). In fact, we can write raw TikZ code directly in **knitr** with the engine `tikz`.

What the `tikz` engine does internally is: use a LATEX template to insert the code chunk and compile the tex document to PDF. By default it uses the template in **knitr** (named tikz2pdf.tex under the misc directory in **knitr**'s installation directory):

```
f <- system.file("misc", "tikz2pdf.tex", package = "knitr")
cat(readLines(f), sep = "\n")

\documentclass{article}
\include{preview}
\usepackage[pdftex,active,tightpage]{preview}
\usepackage{amsmath}
\usepackage{tikz}
\usetikzlibrary{matrix}
\begin{document}
\begin{preview}
%% TIKZ_CODE %%
\end{preview}
\end{document}
```

The line %% TIKZ_CODE %% will be replaced by the TikZ code chunk. If the default template is not satisfactory, we can provide a template via the chunk option engine.opts, e.g., `engine.opts = list(template = 'path/to/tikz/template.tex')`. Then this TEX file is compiled to PDF via the R function `tools::texi2pdf()`. If the specified figure file extension (chunk option `fig.ext`) is not pdf, ImageMagick (via its `convert` utility) will be called to convert the PDF file to other file formats such as PNG, e.g., when the document format is HTML.

Figure 11.1 is a diagram drawn from raw TikZ code below:

```
\usetikzlibrary{arrows}
\begin{tikzpicture}[node distance=2cm, auto,>=latex', thick]
\node (P) {$P$};
```

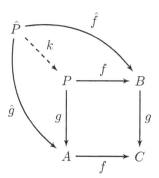

FIGURE 11.1: A diagram drawn with TikZ: the source code is written into a *.tex file and compiled to PDF by LaTeX.

```
\node (B) [right of=P] {$B$};
\node (A) [below of=P] {$A$};
\node (C) [below of=B] {$C$};
\node (P1) [node distance=1.4cm, left of=P, above of=P]
        {$\hat{P}$};
\draw[->] (P) to node {$f$} (B);
\draw[->] (P) to node [swap] {$g$} (A);
\draw[->] (A) to node [swap] {$f$} (C);
\draw[->] (B) to node {$g$} (C);
\draw[->, bend right] (P1) to node [swap] {$\hat{g}$} (A);
\draw[->, bend left] (P1) to node {$\hat{f}$} (B);
\draw[->, dashed] (P1) to node {$k$} (P);
\end{tikzpicture}
```

To develop tikz graphics, the programs `qtikz` or `ktikz` can be help-ful, since they provide a graphical user interface (an editor), which al-lows one to preview the results.

### 11.2.6  Graphviz

Graphviz (Ellson et al., 2002) is an open source and popular graph visu-alization software package (http://www.graphviz.org); it is powerful for drawing diagrams of abstract graphs and networks. Graphviz con-tains a few "filters," such as dot, to draw directed graphs, and neato to draw undirected graphs. When engine = 'dot', dot is used by de-fault; to use other filters, we can set, e.g., engine.path = 'neato'.

Figure 11.2 is an example taken from the documentation of Graphviz.

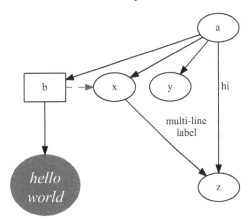

FIGURE 11.2: A diagram drawn with dot in Graphviz (taken from the dot manual).

We used fig.ext = 'pdf' here to produce a PDF graph file, and we can change it to other file formats like PNG as well.

```
digraph test123 {
  a -> b -> c;
  a -> {x y};
  b [shape=box];
  c [label="hello\nworld",color=blue,fontsize=24,
    fontname="Palatino-Italic",fontcolor=red,style=filled];
  a -> z [label="hi", weight=100];
  x -> z [label="multi-line\nlabel"];
  edge [style=dashed,color=red];
  b -> x;
  {rank=same; b x}
}
```

If you want to draw diagrams in HTML documents generated from R Markdown, you may consider the **DiagrammeR** package (https://github.com/rich-iannone/DiagrammeR), which is an HTML widget package that wraps a few JavaScript libraries (see Section 14.5.3 for more information about HTML widgets).

### 11.2.7  Highlight

Highlight is a free and open source software package by Andre Simon (http://www.andre-simon.de) to do syntax highlighting for a large variety of languages, including C, PHP, and R, etc. It can write the output in either LaTeX or HTML.

When the chunk option `engine = 'highlight'`, the `highlight` program is called to generate the highlighted code chunk. The chunk option `engine.opts` is a character string to pass additional arguments to Highlight, e.g., we can specify the input syntax via -S, and the type of output via -O.

The chunk below was taken from the previous awk example; it uses the chunk option `engine.opts = '-S awk -O latex'` to tell Highlight that the input syntax is awk, and the output type is LaTeX, so that Highlight can produce appropriate LaTeX commands on keywords. It may be difficult to see the colors in the printed version of this book, but at least we can see the first line is italic (comments).

```
# how many non-empty lines in the NEWS file?
NF {
    i = i + 1
}
END { print i }
```

Note that Highlight generates commands like `\hlnum{}` (for numbers) and `\hlstr{}` (for strings) to mark up different tokens in the code. These commands are mostly consistent with **knitr**'s syntax highlighting commands, but there are a few exceptions, e.g., `\hlslc{}` (for comments) produced by Highlight is not a part of **knitr**'s commands, so we need to define it in the LaTeX preamble. Similarly, if the Highlight output is HTML, we need to define CSS styles for the class `hl slc`.

## 11.2.8 Other Engines

There are two more engines that are essentially for any language: `cat` and `asis`. The `cat` engine calls the function *cat()* to write the code chunk to a file, and the filename can be provided in the chunk option `engine.opts = list(file = ?)`. The `asis` engine does nothing but just write the code chunk as-is in the output. However, it respects the chunk options `eval` and `echo`: if either of these options is `FALSE`, the code chunk will be hidden from the output, which can be useful when you want to dynamically control whether to show some content in the output.

For example, we can write the code chunk below to a file named styles.css through the `cat` engine:

```
<<engine='cat', engine.opts = list(file = 'styles.css')>>=

p {
  margin: 5px 2px 5px 2px;
}

@
```

The following code chunk will be included in the final output if the variable `internal.only` is TRUE (imagine you have a portion of the report content that you only want to show internally in your group):

```
<<engine='asis', echo = internal.only>>=

Here are some top secrets about our analysis that are hidden
in the public version of this report by setting
'internal.only' to TRUE.

Secret number one: ...

@
```

## 11.3  Persistent Sessions

In fact, there is a major flaw in the engines for interpreted languages introduced before: a new engine session is established for every single code chunk of this engine. This means all code chunks are independent in memory, and the variables created in previous chunks will not be available in latter chunks. The only exception is R code chunks: all of them are evaluated in the same R session. To address this issue, we need to open a persistent session for an engine, and keep on running code chunks in this session. For example, we can create a variable in a Python code chunk, and continue using it in the next Python chunk.

The **runr** package (Xie, 2013) is an attempt to solve this problem. Currently it has experimental support for Bash and Julia code, based on socket connections. The basic idea is like this (take the Julia engine as example):

1. Open a background Julia process that starts a socket server and keeps listening (the background process is detached from

2. the current R session by `system('julia script.jl', wait = FALSE)`);

2. R connects to the Julia socket server via `socketConnection(open = 'w')`, and writes the Julia code chunk to the server;

3. Julia receives the code, evaluates it, and writes the standard output (as plain text) to the socket;

4. R reads from the socket via `socketConnection(open = 'r')`, and writes the Julia output to the report just like R code chunk output;

5. Repeat steps 2–4 if the next Julia code chunk comes in, and Julia will quit if we send the code `quit()` to it.

In this way, the Julia session will be live until we explicitly shut it down from R, and all Julia code chunks will be evaluated in the same Julia session. The **runr** package is still at its early stage, and community contribution is welcome.

# 12

## Tricks and Solutions

In this chapter we show some tricks that can be useful for writing and compiling reports more easily and quickly, and also solutions to frequently asked questions.

## 12.1 Chunk Options

There are a number of built-in chunk options in **knitr**, and we usually assign values to them in chunk headers, but it is still possible to customize these fixed options, e.g., rename the options.

### 12.1.1 Option Aliases

We may feel some options are very frequently used but the names are too long to type. In this case we can set up aliases for chunk options using the function *set_alias()* in the beginning of a document, e.g.,

```
set_alias(w = "fig.width", h = "fig.height")
```

Then we will be able use w and h for the figure width and height, respectively, e.g.,

```
<<fig-size, w=5, h=3>>=
plot(1:10)
@
```

The chunk above is equivalent to:

```
<<fig-size, fig.width=5, fig.height=3>>=
plot(1:10)
@
```

## 12.1.2   Option Templates

Besides option names, we can also bundle frequently used option values together as option templates. The object `opts_template` in **knitr** can be used to build such templates. A template is a named collection of option sets. For example, if there are a large number of plots for which we want to set the graphical device size to be $7 \times 5$ inches, and for other plots, we want the size to be $3.5 \times 3$ inches. We can certainly type `fig.width = 7`, `fig.height = 5` for the first group of plots, and `fig.width = 3.5`, `fig.height = 3` for the second group, but this is apparently tedious (even with option aliases). In this case we can just put the two sets of options in templates:

```
opts_template$set(
    fig.large = list(fig.width = 7, fig.height = 5),
    fig.small = list(fig.width = 3.5, fig.height = 3)
)
```

After the templates have been set up, we can simply use the chunk option `opts.label` in future chunk headers to reference to them. For instance, we want the options for large plots in the chunk below:

```
<<fig-ex, opts.label='fig.large'>>=
plot(1:10)
@
```

This is equivalent to:

```
<<fig-ex, fig.width=7, fig.height=7>>=
plot(1:10)
@
```

## 12.1.3   Program Chunk Options

Since chunk options can take arbitrary R expressions, we can program chunk options besides setting fixed values like numbers or logical values. We show below an example of drawing a table with the **gridExtra** package. First we use the *tableGrob()* function to create a table Grob (graphical object):

```
library(gridExtra)
g <- tableGrob(head(iris))
```

|   | Sepal.Length | Sepal.Width | Petal.Length | Petal.Width | Species |
|---|---|---|---|---|---|
| 1 | 5.1 | 3.5 | 1.4 | 0.2 | setosa |
| 2 | 4.9 | 3.0 | 1.4 | 0.2 | setosa |
| 3 | 4.7 | 3.2 | 1.3 | 0.2 | setosa |
| 4 | 4.6 | 3.1 | 1.5 | 0.2 | setosa |
| 5 | 5.0 | 3.6 | 1.4 | 0.2 | setosa |
| 6 | 5.4 | 3.9 | 1.7 | 0.4 | setosa |

FIGURE 12.1: A table created by the **gridExtra** package: we create a table Grob and draw it in a proper graphical device.

Next, we use *grid.draw()* in the **grid** package to draw the object to a plot. Prior to that, we need to determine an appropriate size for the graphical device; otherwise we might get extra white margins in the plot. In fact, the *convertWidth()* and *convertHeight()* functions in the **grid** package can convert the pre-calculated width and height of the Grob to inches. Therefore, we pass two function calls to the chunk options `fig.width` and `fig.height` instead of using fixed numbers as we usually do. Figure 12.1 is a table of the first four lines of the `iris` data drawn by *grid.draw()*.

```
<<table, fig.width=convertWidth(grobWidth(g), 'in', TRUE)>>=
## width and height in inches
convertWidth(grobWidth(g), "in", value = TRUE)

## [1] 5.55

convertHeight(grobHeight(g), "in", value = TRUE)

## [1] 1.94

grid.draw(g)
@
```

The programmable chunk options enable us to program our reports in many aspects. As one potential application, we may build a linear regression report including common diagnostic procedures, with each procedure in a child document (Section 9.3). Then we can decide whether to include certain procedures based on certain conditions, e.g., if we have detected outliers in the regression model, we include an outlier module to deal with outliers. The chunk below shows a sketch of this idea:

```
<<cooks-distance>>=
cookd <- cooks.distance(fit)
# include an outlier procedure if any distance is
# greater than 1
<<outlier, child=if (any(cookd > 1)) 'outlier.Rnw'>>=
@
```

### 12.1.4   Code in Appendix

Sometimes we do not want to show the code chunks in the body of
the report, but we do not want to completely hide the code, either. In
this case we can move all code chunks to the appendix, and the chunk
option ref.label can be useful here (Section 9.1.2).

   If there are only a small number of code chunks in the document,
we can manually type their labels, e.g.,

```
<<A, echo=FALSE>>=
1+1
<<B, echo=FALSE>>=
2+2
<<C, echo=FALSE>>=
rnorm(10)
<<show-code, ref.label=c('A', 'B', 'C'), eval=FALSE>>=
@
```

   Here we hide the code in the previous chunks by echo = FALSE,
and gather them into the last chunk by ref.label. Note the last chunk
used the chunk option eval = FALSE so that the code is not evaluated
again.

   If there are a lot of code chunks in a document, we can use the function *all_labels()* in **knitr** to obtain all chunk labels in a document, and
pass them to ref.label, e.g.,

```
<<show-code, ref.label=all_labels()>>=
@
```

   We can set echo = FALSE globally by opts_chunk$set(), and use
echo = TRUE for the last chunk to show the code there. Of course we
can also select chunk labels to include there, e.g., remove the first chunk
by all_labels()[-1].

### 12.1.5 Local R Options

The chunk option R.options can take a list of R options to be passed to *options()* for a code chunk. These options will be applied to the code chunk, and restored after the chunk, so it can be useful if you want to temporarily change R options for a particular code chunk.

For example, we use local options width = 30 (the approximate width for printing) and digits = 2 (the number of digits for printing) for the following code chunk:

```
<<R.options = list(width=30, digits=2)>>=

seq(0, 10, length = 20)

## [1] 0.00 0.53 1.05 1.58
## [5] 2.11 2.63 3.16 3.68
## [9] 4.21 4.74 5.26 5.79
## [13] 6.32 6.84 7.37 7.89
## [17] 8.42 8.95 9.47 10.00

@
```

### 12.1.6 Dynamic Code

Usually we just type the code in a chunk, or include code from other chunks by references (Chapter 9). There is yet another way to assign code to a chunk, using the chunk option named code. This makes it possible to construct a code chunk dynamically. For example, you can read the code from an external script:

```
<<code = readLines('foo.R')>>=
@
```

## 12.2 Package Options

Although we did not specifically mention it before, there is an object named opts_knit in **knitr** that controls some package-level options, and its usage is the same as chunk options (opts_chunk).

By default we see a progress bar when we call **knitr**, and we can suppress it by setting opts_knit$set(progress = FALSE). The progress bar shows the progress of *knit()* so we know which chunk is currently being compiled if it takes a relatively long time. To see more information about chunks such as the source code, we can turn on the verbose mode by `opts_knit$set(verbose = TRUE)`.

The package option `root.dir` can be used to set the root working directory when evaluating code chunks. The default working directory is the directory of the input document, but we can change it with this option, e.g., after we set

```
opts_knit$set(root.dir = "/home/foo/bar/")
```

Then we can read a data file under that directory without using the full path, but in general, we recommend putting datasets and source documents in the same directory, and use this directory as the working directory.

For the chunks that are not labeled, automatic labels of the form `unnamed-chunk-`*i* will be used. This can be customized via the package option `unnamed.chunk.label`, e.g.,

```
opts_knit$set(unnamed.chunk.label = "fig")
```

Then the automatic chunk labels will be `fig-1`, `fig-2`, and so on.

## 12.3 Typesetting

In this section we show some solutions to tweaking the typesetting of a report.

### 12.3.1 Output Width

A common problem of using **knitr** in LaTeX is that the output width may exceed the page margin. There are three types of widths: the width of the source code, the text output, and the graphics output. In Section 7.4 we mentioned `\maxwidth`, which guarantees the graphics output will not be wider than the page width.

For the width of source code and text output, it is controlled by the global option `width` in *options()* (Section 6.2.2). The default value for this option is 75, which may be too large for LaTeX documents unless we have reset the page margins (e.g., using the **geometry** package).

When we see the source code or the text output is too wide, we can use a smaller `width` option, e.g.,

```
options(width = 55)
```

However, this may not work all the time: for the source code, R may not be able to find an appropriate place to break the source lines; for text output, the original lines may not contain line breaks (because they are in the verbatim environments, LaTeX will not break the lines automatically). For the example below, the text lines will not be wrapped no matter how small the width option is:

```
# unable to wrap the source code
x <- "thisistoolongandRisunabletofindaplacetoinsertthelinebreak"
# unable to wrap the output line
cat(x, "----")
```

```
## thisistoolongandRisunabletofindaplacetoinsertthelinebreak ---
```

This is an extreme example. Normally our source code can be formatted into several lines. If we have a character string that is too long in the source code, we can consider breaking it into smaller pieces manually and pasting them together with *paste()*, e.g.,

```
x <- paste("this", "is", "too", "long", "and", "R", "is",
    "unable", "to", "find", "a", "place", "to", "insert",
    "the", "line", "break", sep = "")
```

An alternative approach is to use the **listings** style (recall Figure 5.2 and the function *render_listings()*). We can set the breaklines option to true for the **listings** package in the LaTeX preamble:

```
\lstset{breaklines=true}
```

See Figure 12.2 for an example of this option in LaTeX.

## 12.3.2 Message Colors

For LaTeX output, there are three colors defined, corresponding to messages, warnings, and errors, respectively:

```
\definecolor{messagecolor}{rgb}{0, 0, 0}
\definecolor{warningcolor}{rgb}{1, 0, 1}
\definecolor{errorcolor}{rgb}{1, 0, 0}
```

By default messages are black, warnings are magenta, and errors are red. We can redefine them using the command \definecolor{} in the LaTeX preamble.

> We can set the `breaklines` option to `true` to wrap long lines.
>
> ```
> print("asdlfjk sadflkj kljsd klwjr klwjre klwjer kljwre kljwer
>     lkjrwee lkwjre lkwjere lkwjer lkwjre lkasdfa afsd afdafsd
>     afddadf adfsadf afdasdf")
> ```
>
> ```
> [1] "asdlfjk sadflkj kljsd klwjr klwjre klwjer kljwre kljwer
>     lkjrwee lkwjre lkwjere lkwjer lkwjre lkasdfa afsd afdafsd
>     afddadf adfsadf afdasdf"
> ```
>
> By comparison, this shows `breaklines=false`:
>
> ```
> print("asdlfjk sadflkj kljsd klwjr klwjre klwjer kljwre kljwer lkjrw
> ```
>
> ```
> [1] "asdlfjk sadflkj kljsd klwjr klwjre klwjer kljwre kljwer
>     lkjrwee lkwjre lkwjere lkwjer lkwjre lkasdfa afsd afdafsd
>     afddadf adfsadf afdasdf"
> ```

FIGURE 12.2: Break long lines with **listings**: we can use the function *render_listings()* in R and `\lstset{breaklines=true}` in LaTeX.

### 12.3.3   Box Padding

As we introduced in Section 6.2.3, the default LaTeX style of **knitr** is based on the **framed** package, and that is why we see shaded boxes underneath all code chunks. If we feel the default padding of the box is too tight, we can reset the length of `\fboxsep{}` by `\setlength`, e.g.,

`\setlength\fboxsep{5mm}`

```
## an intentional comment to to to to to to to to to to to to
## reach the page margin
rpois(40, 5)
```

```
##  [1]  6  4  6  4  9  5  2  4  2  4  4 10  6  3  1  8  8
## [18]  2  7  4 10  6  5  2  7  4  6  4  2  5  8  7  2  3
## [35]  2  7  7  3  3  3
```

Now we see the gray box is larger, with a padding space of 5 mm. For HTML output, it is much easier to design the style, e.g., we can define the class chunk in CSS as this to make the padding 5 mm:

```
div.chunk {
  padding: 5mm;
}
```

```
\documentclass{beamer}
\begin{document}
\title{Using knitr in Beamer}
\author{Yihui Xie}

\maketitle

\begin{frame}
\frametitle{Introduction}
This is a normal slide.
\end{frame}

% need the option [fragile] for code output!
\begin{frame}[fragile]
\frametitle{Code chunks}
<<test, out.width='.6\\linewidth', fig.align='center'>>=
par(mar = c(4, 4, .1, .1))
x = rnorm(100)
hist(x, main='', col='lightblue', border='white')
rug(x)
@
\end{frame}

\end{document}
```

FIGURE 12.3: A simple example of using **knitr** in beamer slides: note that we need the option [fragile] after \begin{frame}.

### 12.3.4 Beamer

Beamer (Tantau et al., 2012) is a popular document class to create slides with LaTeX. Using **knitr** in beamer slides is not very different from other LaTeX documents; the only thing to keep in mind is that we need to specify the fragile option on beamer frames when we have verbatim output. See Figure 12.3 for the Rnw source of a simple beamer example, with one page of the output in Figure 12.4.

Due to the limited space in beamer slides, it may be desirable to use smaller font sizes for the code. In this case we can set a global chunk option size, e.g.,

Code chunks

```
par(mar = c(4, 4, 0.1, 0.1))
x = rnorm(100)
hist(x, main = "", col = "lightblue", border = "white")
rug(x)
```

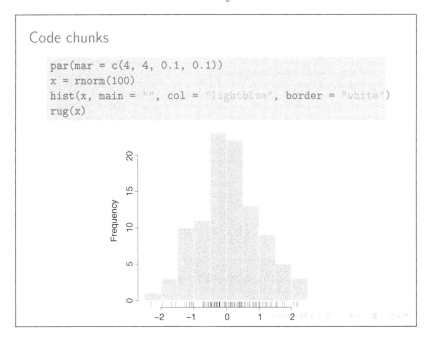

FIGURE 12.4: A sample page of beamer slides: a code chunk with a plot.

```
<<setup, include=FALSE>>=
opts_chunk$set(size = "footnotesize")
@
```

Next we show an example of programming the content of output, which makes it possible to use the beamer command \only{} to show plots one by one in the same place on the screen (for more information, see the beamer manual). The basic idea is to replace the graphics command \includegraphics{} by \only<n>{\includegraphics{}}, with n being the *n*-th plot in the current chunk. Below is a modified plot hook that does this job:

```
<<setup, include=FALSE>>=
hook_plot <- knit_hooks$get("plot")  # the default hook
# tweak and reset the default hook
knit_hooks$set(plot = function(x, options) {
    txt <- hook_plot(x, options)
    if (options$fig.cur <= 0)
        return(txt)
```

```
#' add \only<n> before \includegraphics
gsub("(\\\\includegraphics[^}]+})",
     sprintf("\\\\only<%d>{\\1}", options$fig.cur),
     txt)
})
@
```

One key here is the option `fig.cur`, which is an internal chunk option (not specified by users) providing the current figure number. The substitution of \includegraphics{} was done through regular expressions. After we have modified the plot hook, the plot commands in LATEX output will be changed accordingly.

## 12.3.5 Suppress Long Output

For those who have read the book "Modern Applied Statistics with S" (MASS) by Venables and Ripley (2002), you may have noticed that the authors omitted parts of the output in the book in several places, because the output will otherwise be too long. For example, the data frame `painters` on page 17 has 54 rows, but only the first 5 rows were shown on that page, and the rest of the rows were omitted (the omission was denoted by ....). We can automate this job by redefining the output hook in **knitr** (Section 5.3), e.g.,

```
# the default output hook
hook_output <- knit_hooks$get("output")
knit_hooks$set(output = function(x, options) {
    # print the first 5 lines by default
    if (is.null(n <- options$out.lines))
        n <- 5
    x <- unlist(stringr::str_split(x, "\n"))
    if (length(x) > n) {
        # truncate the output
        x <- c(head(x, n), "....\n")
    }
    # paste first n lines together
    x <- paste(x, collapse = "\n")
    hook_output(x, options)
})
```

Then we can achieve a similar effect of the example in the MASS book:

```
library(MASS)
painters
```

```
##                Composition Drawing Colour Expression
## Da Udine               10       8     16          3
## Da Vinci               15      16      4         14
## Del Piombo              8      13     16          7
## Del Sarto              12      16      9          8
. . . .
```

The basic idea of the hook defined above is, if the number of lines of the output is greater than 5, we extract the first 5 lines by `head(x, 5)`, and append `....` to the output vector, then pass the modified output to the default output hook function *hook_output()*, which was obtained before we reset the output hook. We do not have to hard-code the number of lines to be 5, so we also check if the chunk option `out.lines` is NULL; if it is not, it is supposed to be a number to specify the number of lines to keep in the output. For example, we print the first 10 lines instead:

```
<<print-painters, out.lines=8>>=
library(MASS)
painters
@
```

Note this hook applies to all document formats (Rnw and Rmd, etc.), because we do not have any document-specific code in the new definition; for different document formats, `knit_hooks$get('output')` will be different as well, hence the new hook is portable.

### 12.3.6  Escape Special Characters

As introduced in Section 5.3, the `inline` hook function is used to write inline results into the output. By default, it writes characters as is, and sometimes we may want to escape special characters in LaTeX or HTML, e.g., an inline R code fragment produces a percentage 30%, and we have to write % as \% in LaTeX, otherwise it means LaTeX comments.

It is unclear whether we should escape special characters or not, e.g., we may generate a LaTeX equation from inline R code, in which case we must not escape special characters such as backslashes. Anyway, if we do want to escape them, we can create a new `inline` hook function, e.g.,

```
# get the default inline hook
hook_inline <- knit_hooks$get("inline")
# build a new inline hook
knit_hooks$set(inline = function(x) {
    if (is.character(x))
        x <- knitr:::escape_latex(x)
    hook_inline(x)
})
```

An internal function *escape_latex()* was used to escape special LaTeX characters, and the escaped text strings will be passed to the default `inline` hook. We only added one step before the default hook function, and all features of the default hook will be preserved, such as automatic scientific notation (Section 6.1).

Similarly, if we are writing an R HTML document instead, we can call the *escape_html()* function.

### 12.3.7 The Example Environment

When writting textbooks or tutorials, it can be useful if we number the R code chunks like theorems and equations. It is easy to define an "Example" environment in the LaTeX preamble, e.g., using the **amsthm** package:

```
\usepackage{amsthm}
\newtheorem{rexample}{R Example}[section]
```

Then we can use this new environment `rexample` in our document:

```
\begin{rexample}
<<test, eval=TRUE>>=
1 + 1
rnorm(10)
@
\end{rexample}
```

In fact, we can automate this job with a chunk hook function, so that we do not have to type the environment again and again. The `rexample` hook below writes the environment automatically for a chunk with a non-NULL chunk option `rexample`:

```
knit_hooks$set(rexample = function(before, options, envir) {
    if (before) {
```

```
        sprintf("\\begin{rexample}\\label{%s}\\hfill{}",
            options$label)
    } else "\\end{rexample}"
})
```

Basically this hook writes \begin{rexample} before a chunk, and \end{rexample} after it. Additionally, it writes a label for the environment so that we can reference it later, and the label is the chunk label. Now we can apply it to a chunk, e.g.,

```
<<test, rexample=TRUE>>=
1 + 1
@
```

Figure 12.5 shows a sample page that used this hook function. We can see the R code chunks are numbered after the section numbers, which is due to the [section] option in the definition of the rexample environment. Because the rexample environments also come with labels, we can use \ref{} for cross references.

It is also possible to create a similar hook for R HTML documents, but since HTML is not primarily for typesetting purposes, it is not easy to get the automatic numbering as in LATEX. Anyway, we can use our own counter in R, e.g.,

```
## an example counter for HTML
example_count <- 0
knit_hooks$set(rexample = function(before, options, envir) {
    if (before) {
        # increment by 1
        example_count <<- example_count + 1
        sprintf("<div>Example %d</div>", example_count)
    } else ""
})
```

### 12.3.8   The Docco Style

Besides LATEX documents, you can also use typeset HTML documents. There is a function *rocco()* in **knitr** that provids a two-column layout for HTML documents. This style was borrowed from a literate programming package named Docco (https://github.com/jashkenas/docco). The narratives and code are arranged in separate columns, so that you can keep on reading either the narratives or the code in one

---

# 1 Introduction

This is a test of the R Example environment.

## 1.1 Go!

**R Example 1.1.**

```
1 + 1

## [1] 2
```

Look at Example 1.1!

## 1.2 Ha!

**R Example 1.2.**

```
x = rnorm(10)
```

Move on!

**R Example 1.3.**

```
sd(x)  # standard deviation

## [1] 1.124
```

---

FIGURE 12.5: R code chunks in the R Example environments: the examples are numbered following the section numbers.

column. You can hide either column with a keyboard shortcut. Figure 12.6 is a screenshot of a package vignette in **knitr** that uses this style:

```
vignette("docco-classic", package = "knitr")
```

## 12.4 Utilities

There are a few utility functions in **knitr** to complete miscellaneous tasks such as writing BibTeX databases for R packages, base64 encoding

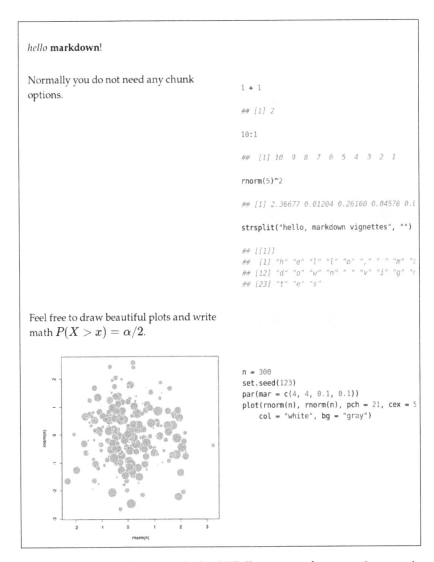

FIGURE 12.6: The Docco style for HTML output: the narratives are in the left column, and the R code is in the right column. You can render such a page from R Markdown using the function *rocco()* in **knitr**.

images for HTML output, and compiling source documents to the final output.

## 12.4.1 R Package Citation

The function *write_bib()* is a wrapper to the functions *citation()* and *to-Bibtex()* in base R. By default it collects the packages loaded into the current R session and extracts their citation information. It also has an argument named `tweak`, which determines whether to tweak the default citation information, e.g., the author name "Duncan Temple Lang" should be "Duncan {Temple Lang}" in the bibliography database. Instead of manually modifying information like this, *write_bib()* can automatically deal with it.

```
write_bib(c("filehash", "RGtk2", "rms"))

@Manual{R-filehash,
  title = {filehash: Simple key-value database},
  author = {Roger D. Peng},
  year = {2014},
  note = {R package version 2.2-2},
  url = {http://CRAN.R-project.org/package=filehash},
}
@Manual{R-RGtk2,
  title = {RGtk2: R bindings for Gtk 2.8.0 and above},
  author = {Michael Lawrence and Duncan {Temple Lang}},
  year = {2014},
  note = {R package version 2.20.31},
  url = {http://CRAN.R-project.org/package=RGtk2},
}
@Manual{R-rms,
  title = {rms: Regression Modeling Strategies},
  author = {Frank E. {Harrell, Jr.}},
  year = {2015},
  note = {R package version 4.3-0},
  url = {http://CRAN.R-project.org/package=rms},
}
```

The second argument of *write_bib()* is `file`, and we can pass a filename to it to save the bibliography items into a file. By default, it writes to the standard output.

The advantage of generating the bibliography database using this function is that we can guarantee we always cite the package versions that we really use in a document. If we hard-code the bibliography, the citations may be out-of-date after we update R packages.

If we do not want to write the file each time we compile the document, we can cache the chunk. Then a natural question is, when should we, or how can we update the cache? Recall Chapter 8 and one solution is to put the package version(s) in a chunk option, e.g., if the main package that we use for a document is called **foo**, we can write a chunk like this:

```
<<write-bib, cache=TRUE, version=packageVersion('foo')>>=
write_bib(c("foo", "other", "packages"), file = "paper.bib")
@
```

Then whenever the **foo** package is updated, the cached chunk will be updated accordingly.

### 12.4.2   Image URI

It is convenient to publish a PDF report because a PDF document contains everything in one file, including plots in particular, but that is not true for HTML reports. If an HTML page contains images that are external files, we have to publish these images along with the HTML file, otherwise the Web browser will not be able to find them. There is a technology called "Data URI" in Web pages that solves this problem. In short, we can encode a file into a character (base64) string and include it in HTML, so that we do not need the original file any more when publishing the HTML page. In other words, the HTML page is self-contained just like PDF.

The function *image_uri()* in **knitr** was designed to encode images as base64 strings. Obviously it only applies to HTML output (including Markdown). We can enable this function in opts_knit:

```
opts_knit$set(upload.fun = image_uri)
```

Then if we have plots in HTML output, the image file paths will be replaced by base64 character strings. Below is an example of encoding the R logo (a JPEG image):

```
# encode the R logo
logo <- file.path(R.home("doc"), "html", "logo.jpg")
uri <- image_uri(logo)
# the first 250 characters
uri.sub <- substring(uri, seq(1, 201, 50), seq(50, 250,
```

```
    50))
cat(uri.sub, sep = "\n")
```

data:image/jpeg;base64,/9j/4AAQSkZJRgABAQEBKwErAAD
/4QAWRXhpZgAATUOAKgAAAAgAAAAAAAD/2wBDAAUDBAQEAwUEB
AQFBQUGBwwIBwcHBw8LCwkMEQ8SEhEPEReTFhwXExQaFRERGCE
YGhOdHx8fExciJCIeJBweHx7/2wBDAQUFBQcGBw4ICA4eFBEUH
h4eHh4eHh4eHh4eHh4eHh4eHh4eHh4eHh4eHh4eHh4eHh4eHh4

### 12.4.3  Upload Images

Based on the same reason, we designed another function *imgur_upload()* to upload images to the website Imgur.com, and this function returns the URL of the uploaded image. Then, instead of using the image file path to reference the image (which has the problem mentioned before), we use a URL that is accessible anywhere as long as we have Internet connection. To continue the previous example, we can upload the R logo to Imgur website by:

```
imgur_upload(logo)
```

This returns a URL of the form `http://i.imgur.com/xxxxx.jpg`. To make things even easier, we can set the package option `upload.fun` like we did in the last section:

```
opts_knit$set(upload.fun = imgur_upload)
```

Then images will be automatically uploaded to Imgur when we knit a document. To avoid repeated uploading of the same image, we can turn on cache.

### 12.4.4  Compile Documents

For some document formats, there are two steps in compilation. For example, Rnw documents are compiled through **knitr** to LaTeX documents, which need to be compiled to PDF via LaTeX. For Rmd documents, the final product is often HTML instead of Markdown, which is the direct output of **knitr**.

To turn the two steps into one, the functions *knit2pdf()* and *knit2html()* can be used. The former will first *knit()* an Rnw document to a TeX document, and then call *texi2pdf()* in base R to compile it to PDF; the latter will *knit()* an Rmd document to a Markdown document, and call *markdownToHTML()* in the **markdown** package to compile Markdown to HTML.

For users under Unix-like systems, there is a Bash script named knit under the directory bin of **knitr**'s installation path; we can find it via:

```
system.file("bin", "knit", package = "knitr")

## [1] "/home/yihui/R/knitr/bin/knit"
```

It is an executable script that calls R to load **knitr** and automatically uses *knit2pdf()* or *knit2html()* based on the filename extension; if we put this script in the PATH variable, we can call it in command line directly. For example, I have made a symbolic link under ~/bin/ to this script, and added this to ~/.bashrc:

```
PATH=$PATH:$HOME/bin
export PATH
```

Then we can run knit like other programs in the terminal without having to start R and type all the commands there.

## 12.4.5   Construct Code Chunks

So far we have been using files as the input for the *knit()* function in **knitr**. As a matter of fact, there is an alternative argument to receive the source document, which is named text.

```
# arguments of knit()
formatR::usage(knit, width = 40)

## knit(input, output = NULL, tangle = FALSE,
##     text = NULL, quiet = FALSE, envir = parent.frame(),
##     encoding = getOption("encoding"))
```

If we provide an input file to *knit()*, it will be read into **knitr** and assigned to the text argument eventually. The content of files is usually fixed, but for the text argument, we can dynamically construct it using R since it is nothing but a character variable.

Now we show a comprehensive example, which builds a PDF document for all the geom examples in the **ggplot2** package; see the source code in Figure 12.7 and a sample page of the output in Figure 12.8. It may look a little bit complicated at first glance, but the basic idea is simple:

1. in the `setup` chunk, we set two global chunk options: `tidy` = `FALSE` (optional) and `cache` = `TRUE` (because there are a large number of example code chunks to run later);

2. in the `write-examples` chunk, we use *apropos()* to find all function names that start with geom_; then we find their help files and from there extract the examples code with *Rd2ex()* in the **tools** package; finally we construct Rnw chunks using the function names as section titles and chunk labels, and assign the source text to a variable `ex`;

3. in the last step, we knit the source passed from the text argument and *knit()* returns the LaTeX code, which we insert into the document as a text string by `\Sexpr{}`;

This source document will produce a PDF document of more than 200 pages, taking a few minutes on the first run. Note that it uses the document class `tufte-handout`, which is a LaTeX class you may have to install (it is not a standard class that comes by default).

## 12.4.6 Extract Source Code

We mentioned the function *purl()* briefly in Section 3.4. Actually it has an additional argument named `documentation`, which controls the level of details of documentation chunks.

```
args(purl)

## function (..., documentation = 1L)
## NULL
```

The `documentation` argument takes three possible values:

**0L** discard all text chunks, including chunk headers, so the output is pure program code

**1L** discard text chunks but preserve chunk headers in the exported code file

**2L** keep everything in the source document but put text chunks in roxygen comments (i.e., after #')

The following chunk shows examples corresponding to three values of the `documentation` argument. Note that the chunk headers are written after ## ----, and text chunks are after #'. When `documentation` = 2, the generated R script can be passed to the function *spin()* to restore the original document (Section 5.4).

```
\documentclass[a4paper,titlepage]{tufte-handout}
\title{ggplot2 Gallery}
\begin{document}
\maketitle
\tableofcontents

<<setup, include=FALSE>>=
# cache chunks and do not tidy ggplot2 examples code
opts_chunk$set(tidy = FALSE, cache = TRUE)
@

% all geoms in ggplot2
<<write-examples, include=FALSE>>=
library(ggplot2)
ex = lapply(apropos("^geom_"), function(g) {
  p = utils:::index.search(g, find.package(), TRUE)
  tools::Rd2ex(utils:::.getHelpFile(p), f <- tempfile())
  c(sprintf("\\section{%s}\n\n<<%s>>=",
            knitr:::escape_latex(g), g),
    readLines(f), "@\n\n")
})
@

\Sexpr{knit(text = unlist(ex))}

\end{document}
```

FIGURE 12.7: The source document of the **ggplot2** geom examples: the *Rd2ex()* function was used to extract all examples code for the geom functions, and we construct code chunks using the Rnw syntax for **knitr** to compile.

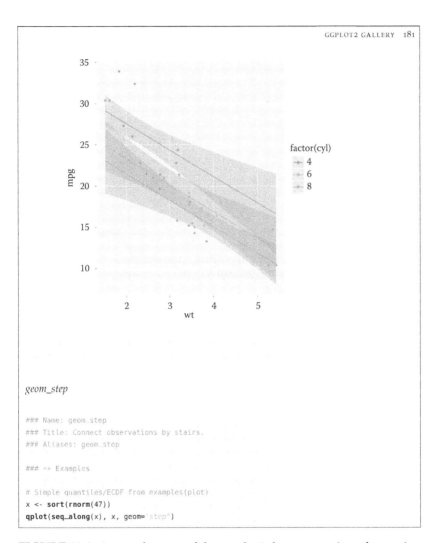

FIGURE 12.8: A sample page of the **ggplot2** documentation: the section titles, code, and plots are all dynamically generated.

```
src <- c("this is the source document", "<<A, tidy=FALSE>>=",
    "1+1", "@", "the end")
cat(purl(text = src, documentation = 0L))
```

```
1+1
```

```
cat(purl(text = src, documentation = 1L))
```

```
## ----A, tidy=FALSE-------------------------------------
1+1
```

```
cat(purl(text = src, documentation = 2L))
```

```
#' this is the source document
## ----A, tidy=FALSE-------------------------------------
1+1
```

```
#' the end
```

For code chunks that have the chunk option `purl = FALSE`, their code will be ignored. For those chunks that have `eval = FALSE`, their code will be commented out.

### 12.4.7   Reproducible Simulation

As we discussed in Chapter 8, it is not trivial to write a report that can be easily and completely reproducible for others. One challenge is to make random simulations reproducible. Of course we can use *set.seed()* to fix the random seed, but what if we have enabled cache?

The problem is, when should we update a cached chunk that involves random numbers? One sufficient condition is the change of the random seed, i.e., if the random seed has changed before a chunk, this chunk should be re-evaluated.

The object `rand_seed` in **knitr** was designed for this purpose. This object is essentially an unevaluated expression:

```
rand_seed
```

```
## {
##     if (exists(".Random.seed", envir = globalenv()))
##         get(".Random.seed", envir = globalenv())
## }
```

```
is.language(rand_seed)
```

```
## [1] TRUE
```

Basically it returns the random seed if it exists. We can assign this object to a chunk option; because it is an unevaluated expression, each time a chunk is compiled, this object will be evaluated again (**knitr** will always evaluate unevaluated chunk options). Then if the random seed has changed, **knitr** will be able to detect the change and update the cached chunk accordingly. Below is an example:

```
<<random-cache, cache=TRUE, cache.extra=rand_seed>>=
x <- rnorm(100)
@
```

Even if we only switched the positions of two cached chunks (with the code and options untouched), the cache will be invalidated because the evaluated results of rand_seed will be different for these two chunks compared to the last run.

### 12.4.8  R Documentation

R has a standard documentation system, and one thing that can be improved is the examples in the help pages — we can actually run these examples and put the results in the pages, so that it is easier for the reader to know the results without having to copy and paste code from the documentation.

The function *knit_rd()* was designed for this task: it takes a package name and extracts all its HTML help pages, then compiles all the examples. This can be handy for package authors, because it generates HTML files that can be published on the Web, and they are richer than the default R documentation. For example, we recompile all the help pages of the **rpart** package:

```
knit_rd("rpart")
```

We will see a few HTML files under the current working directory. If there are plots in the examples, they will be base64 encoded and embedded in the pages, so we do not need to take care of additional files — just upload all these HTML files to a website.

### 12.4.9  Rst2pdf

Rst2pdf (http://rst2pdf.ralsina.com.ar) is a free software package to create PDF from reStructuredText. If we write the source document

```
\documentclass{article}
\begin{document}
<<read-demo>>=
library(diagram)
read_demo('flowchart', package = 'diagram',
          labels = 'demo-flowchart')
<<demo-flowchart, dev='tikz', cache=TRUE>>=
@
\end{document}
```

FIGURE 12.9: The `flowchart` demo in the **diagram** package: we read the demo into **knitr**, assign a label `demo-flowchart` to it, and insert it into the document using this label.

in the R reST format (Section 5.2.4), the output from **knitr** is a *.rst document, and we can call Rst2pdf (if installed) to convert it to PDF via the wrapper function *rst2pdf()* in **knitr**, or just call `knit2pdf('foo.Rrst')` in one step.

### 12.4.10   Package Demos

Some R packages contain demos, which can be run by the *demo()* function, e.g.,

```
demo("plotmath")
demo("notebook", package = "knitr")
```

We can insert demos into a source document using the *read_demo()* function in **knitr**, which is simply a wrapper of *read_chunk()* as introduced in Section 9.2.2.

Figure 12.9 shows a complete example of including the `flowchart` demo of the **diagram** package into an Rnw document; see Figure 12.10 for a sample page of the output. We can certainly use a simple chunk of one line of code `demo('flowchart', echo = TRUE)` instead, but we will lose syntax highlighting.

### 12.4.11   Pretty Printing

When we want to see the source code of an R function, we can simply type its name and R will print its source code, e.g.,

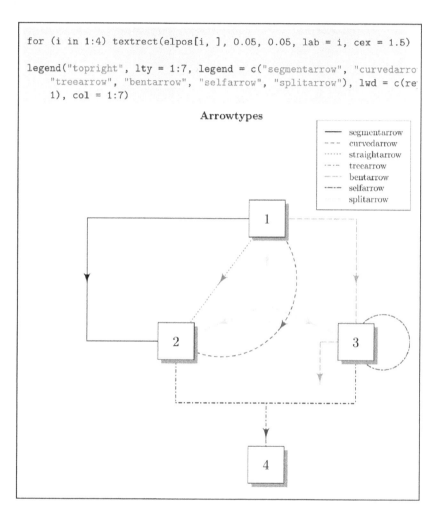

FIGURE 12.10: A sample page of the flowchart demo: we can see the syntax highlighting as well as the diagram.

```
fivenum
```

```
## function (x, na.rm = TRUE)
## {
##     xna <- is.na(x)
##     if (any(xna)) {
##         if (na.rm)
##             x <- x[!xna]
##         else return(rep.int(NA, 5))
##     }
##     x <- sort(x)
##     n <- length(x)
##     if (n == 0)
##         rep.int(NA, 5)
##     else {
##         n4 <- floor((n + 3)/2)/2
##         d <- c(1, n4, (n + 1)/2, n + 1 - n4, n)
##         0.5 * (x[floor(d)] + x[ceiling(d)])
##     }
## }
## <environment: namespace:stats>
```

But since **knitr** supports syntax highlighting and code reformatting (Sections 6.2.2 and 6.2.3), we may also want to use these features on the function source. The only question is how to get the source code into **knitr**, and one answer could be *read_chunk()* again. We define a function *insert_fun()* below to assign the (dumped) source code of an R object to a chunk:

```
insert_fun <- function(name) {
    read_chunk(lines = capture.output(dump(name, "")),
        labels = paste(name, "source", sep = "-"))
}
```

For an object name, its dumped representation will be captured in a code chunk of the label name-source (see ?dump and ?capture.output for details). Now we can use this function to insert the source code of any functions into the source document, e.g., the *fivenum()* function:

```
insert_fun("fivenum")
```

Then we only need to use the chunk label fivenum-source to show the (highlighted and reformatted) source code:

```
fivenum <- function(x, na.rm = TRUE) {
    xna <- is.na(x)
    if (any(xna)) {
        if (na.rm)
            x <- x[!xna] else return(rep.int(NA, 5))
    }
    x <- sort(x)
    n <- length(x)
    if (n == 0)
        rep.int(NA, 5) else {
        n4 <- floor((n + 3)/2)/2
        d <- c(1, n4, (n + 1)/2, n + 1 - n4, n)
        0.5 * (x[floor(d)] + x[ceiling(d)])
    }
}
```

The source code of the above chunk is:

```
<<fivenum-source>>=
@
```

## 12.4.12 A Macro Preprocessor

The function *knit_expand()* was designed to pre-process a source document, which is often a template file for creating repeated text with some changing parameters. For example, we may want to build regression models for the same response variable against different independent variables, and all the models are more or less the same form; all we need to change is the variable names in the models. For example, linear regressions of mpg against two variables in the mtcars data:

```
fit1 <- lm(mpg ~ cyl + disp, data = mtcars)
fit2 <- lm(mpg ~ hp + drat, data = mtcars)
```

The basic idea of *knit_expand()* is to insert some tags in a template, and dynamically evaluate them in the current environment. Below are a few simple examples:

```
knit_expand(text = "The value of pi is {{ round(pi,4) }}.")

## [1] "The value of pi is 3.1416."
```

```
knit_expand(text = "The value of pi is {{ round(pi,4) }}.",
    pi = 1.234567)
```

```
## [1] "The value of pi is 1.2346."
```

```
knit_expand(text = "radius = {{r}} and area = {{pi*r^2}}",
    r = 5)
```

```
## [1] "radius = 5 and area = 78.5398163397448"
```

```
knit_expand(text = "$a = {{a}}$ and $b = {{b}}$", a = 1,
    b = 2)
```

```
## [1] "$a = 1$ and $b = 2$"
```

As we can see above, the R expressions in {{}} are evaluated and their values are written in the output.

We can dynamically create the source document for *knit()* based on *knit_expand()* like the example in Section 12.4.5. As an example, we build the linear regression models of mpg against all combinations of two variables in the mtcars data, with each model in one section. We write a template file as shown in Figure 12.11 and name it mtcars-template.Rnw. Then we can build our models based on this template:

```
## we can build one model of mpg vs cyl+disp by
knit_expand("mtcars-template.Rnw", x1 = "cyl", x2 = "disp",
    i = 1)
## and we can vectorize the whole job with mapply()
vars <- combn(names(mtcars)[-1], 2)
src <- mapply(knit_expand, file = "mtcars-template.Rnw",
    x1 = vars[1, ], x2 = vars[2, ], i = seq_len(ncol(vars)))
```

We used the function *combn()* to get all combinations of two variables, and passed them to *knit_expand()* via *mapply()*. The next step is straightforward: pass the pre-processed source text src to *knit()*, e.g., knit(text = src, output = 'lm-mtcars.tex'), and we will get the LaTeX output with the regression results.

## 12.4.13  Exit Knitting Early

Sometimes you may not want to knit the whole document, and the function *knit_exit()* allows you to quit early. Once you put it in a code chunk, the rest of the document will be ignored, and the results from all previous text/code chunks will be returned immediately.

```
\section{Regression against {{x1}} and {{x2}}}

<<lm-{{x1}}-{{x2}}>>=
fit{{i}} = lm(mpg ~ {{x1}} + {{x2}}, data = mtcars)
summary(fit{{i}})
@
```

FIGURE 12.11: A template of regression models: the variables x1 and x2 will be substituted by two variable names in mtcars, the chunk labels are also created from variable names (so they are unique).

## 12.4.14 Literal knitr Source Code

You may find it a difficult task when you want to write literal **knitr** source code, such as the source code of an inline R expression, e.g., \Sexpr{x}. This is a common task especially when you write **knitr** tutorials. You certainly cannot write the source code as-is, because **knitr** will evaluate it. You cannot even write \verb|\Sexpr{x}|, since **knitr** does not understand the special meaning of the LATEX command \verb||. Similarly, it may be difficult to write a literal inline expression `r x` in R Markdown.

The function *inline_expr()* in **knitr** provides one solution to this problem. It takes a character string, and wraps it using the appropriate syntax of inline expressions.

```
inline_expr("1 + 1")

## [1] "\\Sexpr{1 + 1}"

inline_expr("paste('a', 'b')")

## [1] "\\Sexpr{paste('a', 'b')}"
```

Then you can call this function in an inline expression. For example, \verb|\Sexpr{inline_expr('1 + 1')}| in Rnw documents, or `` `r inline_expr{'1 + 1'}` `` in Rmd documents.

Another solution is to mutate certain characters in the inline expression, e.g., instead of \Sexpr{}, you can write \textbackslash{}Sexpr{} in LATEX, since the latter will not be recognized as an inline expression.

There is a similar challenge for writing literal code chunks. Again, you just need to change the source code of the code chunk so that it is no longer recognizable by **knitr**. For example, you can add an inline expression with an empty character string before the chunk header, such as \Sexpr{"}<<>>=, or `` `r ''`````{r}``. Such lines will not be treated as valid chunk headers, because **knitr**'s syntax only allows white spaces before the chunk header.

## 12.4.15 Spell Checking

Base R has a spell check function *aspell()* in the **utils** package, which can perform spell check via Aspell, Hunspell, or Ispell. To check the spelling of **knitr** documents, you may want to skip code chunks, because program code often contains words that are considered as misspelled.

The *aspell()* function can take a filter function to skip certain lines in the files. The function *knit_filter()* was designed to skip code chunks in a file. Here are two examples of checking an Rnw and Rmd file, respectively:

```
library(knitr)
knitr_example <- function(...) system.file("examples", ...,
    package = "knitr")

# -t means the TeX mode
aspell(knitr_example("knitr-minimal.Rnw"), knit_filter,
    control = "-t")

## backref
##    /home/yihui/R/knitr/examples/knitr-minimal.Rnw:13:37
##
## boxplots
##    /home/yihui/R/knitr/examples/knitr-minimal.Rnw:41:45
##
## colorlinks
##    /home/yihui/R/knitr/examples/knitr-minimal.Rnw:13:51
##
## knitr
##    /home/yihui/R/knitr/examples/knitr-minimal.Rnw:26:26
....

# -H is the HTML mode
aspell(knitr_example("knitr-minimal.Rmd"), knit_filter,
    control = "-H -t")

## knitr
```

```
##   /home/yihui/R/knitr/examples/knitr-minimal.Rmd:3:38
##   /home/yihui/R/knitr/examples/knitr-minimal.Rmd:59:42
##
## LaTeX
##   /home/yihui/R/knitr/examples/knitr-minimal.Rmd:38:1
```

You can add words that you know are correctly spelled to a dictionary, so the spell checker does not report them the next time. R has a built-in dictionary, which contains the word "LATEX". Once we apply this dictionary, you will see the word "LATEX" is no longer reported (but "knitr" still is):

```
# use a dictionary: LaTeX is a known word
dict <- Sys.glob(file.path(R.home("share"), "dictionaries",
    "*.rds"))
# what's in the dictionary?
if (length(dict) >= 1) head(readRDS(dict[1]), 20)
```

```
##  [1] "Accessor"          "accessor"
##  [3] "accessors"         "ACF"
##  [5] "Affymetrix"        "AIC"
##  [7] "Akaike"            "Akaike's"
##  [9] "alikes"            "ANOVA"
## [11] "API"               "approximative"
## [13] "ARIMA"             "ARMA"
## [15] "ascii"             "AUC"
## [17] "autocorrelation"   "autocorrelations"
## [19] "autocovariance"    "autocovariances"
```

```
aspell(knitr_example("knitr-minimal.Rmd"), knit_filter,
    control = "-H -t", dictionaries = dict)
```

```
## knitr
##   /home/yihui/R/knitr/examples/knitr-minimal.Rmd:3:38
##   /home/yihui/R/knitr/examples/knitr-minimal.Rmd:59:42
```

## 12.5 Debugging

Although there is no hard requirement on whether to run **knitr** in an interactive or non-interactive R session, it is recommended to use a new non-interactive R session because it is less likely to be "polluted" by existing objects in the R workspace. Based on this consideration, some editors such as RStudio open a new R session to compile reports by default.

The problem with non-interactive R sessions is that debugging may be inconvenient. If an error occurs, **knitr** will quit from R with a message printed on screen showing the problematic chunk, including its label and line numbers.

If the information mentioned above is not enough, we can also open an interactive R session and run *knit()* there. When an error occurs in this case, we can use common debugging tools such as *traceback()* (to see the call stacks that led to the error), or *debug()*, or *browser()*.

## 12.6   Multilingual Support

If the source document was not encoded with the native encoding of the current system, we will have to manually specify its encoding via the encoding argument in *knit()*. For example, if the source document was written in Simplified Chinese and encoded in GB2312, we need to compile it by:

```
knit("yourfile.Rnw", encoding = "GB2312")
```

Note that **knitr** does not try to automatically detect the encoding of the input document, but the editors usually know the encoding information about the documents. For example, both RStudio and LyX will pass the encoding string to **knitr** before a document is compiled.

# 13

## *Publishing Reports*

After compiling a report through **knitr**, the output document may not be the end product directly. In particular, output from Rnw documents and Rmd documents often needs further compilation. The direct output from Rnw is LaTeX, which can be compiled to PDF. The output from Rmd is Markdown, and what we really read is a Web page converted from Markdown.

There is not much left to do with LaTeX — the tool chain is fairly standard and mature (LaTeX, PDFTeX, XeTeX, and LuaTeX, etc). When we publish reports based on Rnw source documents, we only need to publish a single PDF file. One thing that we may need to do is to hide the source code, since the reader may not be interested in reading it. In that case, we can set the chunk option echo to be FALSE globally, and sometimes we may also want to hide the messages and warnings from R:

```
<<setup, include=FALSE>>=
knitr::opts_chunk$set(
  echo = FALSE, message = FALSE, warning = FALSE
)
@
```

Then only the results will be shown in the final report. In this chapter, we introduce some tools that can help us convert the results from **knitr** to end products, as well as some presentation tools.

## 13.1 RStudio

As we have introduced in Section 4.1, RStudio has comprehensive support for **knitr**. One thing that RStudio has made really easy is the publishing of HTML reports produced from R Markdown. After we click the Knit HTML button, we can see a button named Publish in the toolbar of the preview page. This button enables us to publish the report to the website `http://rpubs.com` with one click. You need to register on the website in advance so that the report can be published to your account.

What happens behind the scenes when we click the Knit HTML button is that RStudio calls **knitr** to compile Rmd to Markdown, then RStudio calls Pandoc to convert Markdown to HTML. In the second step, Pandoc tries to find out all possible images in the document and encodes them as base64 strings (Section 12.4.2) so that the HTML file becomes self-contained. When we publish them to the website, we do not need to upload image files separately. Alternatively, we can use *imgur_upload()* introduced in Section 12.4.3 to upload images to Imgur.

Besides encoding images, Pandoc also detects LaTeX math expressions in the document; if there are any, the JavaScript library MathJax will be used in the HTML header, so that math expressions are rendered correctly on the Web page.

## 13.2 Pandoc

Pandoc (`http://johnmacfarlane.net/pandoc`) is a universal document converter. In particular, Pandoc can convert Markdown to many other document formats, including LaTeX, HTML, Rich Text Format (*.rtf), E-Book (*.epub), Microsoft Word (*.docx), and OpenDocument Text (*.odt), etc. This section tells you how Pandoc works under the hood, and you should see Chapter 14 for R Markdown v2, which is much more convenient to work with than what we introduce in this section.

Pandoc is a command line tool. Linux and Mac users should be fine with it; for Windows users, the command window can be accessed via the Start menu, then Run `cmd`. Once we have opened a command window (or terminal), we can type commands like this to convert a Markdown file, say, test.md, to other formats:

```
pandoc test.md -o test.html
pandoc test.md -s --mathjax -o test.html
pandoc test.md -o test.odt
pandoc test.md -o test.rtf
pandoc test.md -o test.docx
pandoc test.md -o test.pdf
pandoc test.md --latex-engine=xelatex -o test.html
pandoc test.md -o test.epub
```

The option -o specifies the output filename. Figure 13.1 shows a screenshot of an OpenDocument Text document, which looks very much like Microsoft Word in terms of the appearance.

There is a function *pandoc()* in **knitr** that calls Pandoc from R. It also enables us to embed Pandoc arguments in Rmd documents; see its documentation for details.

It is always a big challenge to find a document format that works universally. Some users are not satisfied with Word, and other users find LaTeX difficult to learn. Markdown can be one possible solution due to Pandoc's support for a large variety of document formats. However, the details in typesetting may not be satisfactory in all document formats, and we are very likely to have to manually tweak the converted documents later.

## 13.3   HTML5 Slides

To make presentations, we can use the Beamer class mentioned in Section 12.3.4. With the development of Web technologies, we can also make HTML slides on the Web, which we can view in Web browsers, instead of having to download the slides as (PDF or PPT) files as usual. HTML5 slides also enable us to embed rich media in slides such as video clips and interactive content (e.g., JavaScript applications).

There are a number of ways to make HTML5 slides. One way is to go from Markdown with Pandoc. Figure 13.2 shows an Rmd document, which can be compiled to Markdown through **knitr**; then we can call Pandoc to convert it to HTML5 slides in the command line (suppose the filename is test.md):

```
pandoc -s -t dzslides test.md -o test.html
```

The option -s tells Pandoc to generate a standalone document (with all CSS definitions written into this document); the option -t means the format to generate to; note that dzslides is only one possible value for HTML5 slides; see the online documentation of Pandoc for other formats.

Now we can open the HTML file in a Web browser and use the left/right arrows to navigate through slides.

If we are uncomfortable with command line tools, there are a few R packages such as **slidify** (Vaidyanathan, 2012) and **rmarkdown** (Allaire et al., 2015a) that can make HTML slides. We can create HTML slides directly from Rmd files, and there are also some nice templates and themes shipped with these packages.

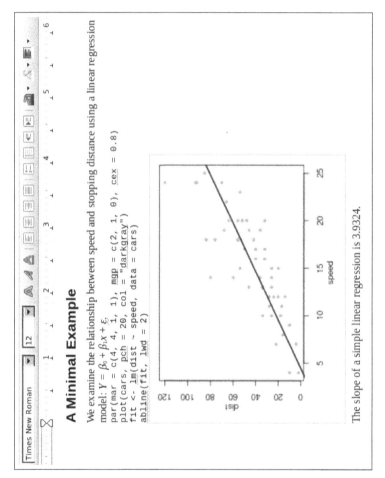

FIGURE 13.1: OpenDocument Text converted from Markdown: we used the same Markdown document in Section 3.2.2 but removed the chunk option `fig.align='center'`.

```
% Writing beautiful and reproducible slides quickly
% Yihui Xie
% 2012/12/05

# Introduction

- knitr
- pandoc

# A code chunk

```{r computing}
head(cars)
cor(cars)
```
```

FIGURE 13.2: The source of an example of HTML5 slides: we can compile this document through **knitr**, then convert the Markdown output to DZSlides via Pandoc.

## 13.4   Jekyll

Jekyll (http://jekyllrb.com) is a blog engine based on plain text files. The blog posts can be written in Markdown, therefore it is possible to publish results from **knitr** to websites. One thing that we need to pay attention to is that the syntax of code blocks is different with traditional Markdown (three backticks): for Jekyll, we need to put code blocks in the Liquid tag:

```
{% highlight lang %}
# code here
{% endhighlight %}
```

We do not need to worry about this technical detail because **knitr** has a renderer for Jekyll: *render_jekyll()*. After we call this function, the R code and its output will be written into the correct tags. Actually the syntax for code blocks also depends on which markdown renderer you use for Jekyll. The default renderer is **kramdown** (http://kramdown.gettalong.org), which does not support three backticks, but some other renderers may support this syntax, such as **redcarpet** (`https://github.com/vmg/redcarpet`). Again, the big trouble of Markdown is that the syntax is different in different renderers, as we have mentioned in Section 5.2.1.

In fact, the website of **knitr** (`http://yihui.name/knitr`) was built with Jekyll and hosted on Github.

## 13.5 WordPress

WordPress is a free, open-source, and popular blogging system based on PHP and MySQL. It has an API that allows one to publish blog posts from a third-party client. The **RWordPress** package provides R functions to communicate with a WordPress site. There is a wrapper function *knit2wp()* in **knitr** that makes it possible to compile an Rmd document and send it to WordPress directly. See `http://yihui.name/knitr/demo/wordpress/` for details of configurations such as the login name and password.

# 14

## R Markdown

There has been a lot of progress on the R Markdown development since the first edition of this book. To make it clear, there are two versions of R Markdown: we call the implementation in the **markdown** package (Allaire et al., 2015b) "R Markdown v1" (`https://github.com/rstudio/markdown`), and we call the implementation **rmarkdown** (Allaire et al., 2015a) "R Markdown v2" (`http://rmarkdown.rstudio.com`). Unless otherwise noted, use of the term "R Markdown" in this chapter refers to R Markdown v2.

R Markdown v1 is based on the C library **sundown**, and the major focus is HTML output. Its functionality is very limited, e.g., there is no support for citations or footnotes. R Markdown v2 is based on Pandoc, which has boosted Markdown to a whole new level. There are two aspects of the improvements: the Pandoc Markdown syntax is richer, so we can write more types of elements, and the output format is no longer limited to HTML — we can also export Markdown to LaTeX/PDF, Word, and HTML5 slides, etc. In this chapter, we will introduce the design philosophy of **rmarkdown**, what it can do, and how to customize or extend it.

## 14.1  Overview

Although **knitr** supports a variety of document formats (Chapter 5), R Markdown is probably the most popular one. Markdown, limited as it is in terms of functionality, is a nice document language for beginners. On the other hand, authors may not even want a lot of features at all. Markdown may be restrictive in the eyes of LaTeX users, but not everyone needs to care that much about typesetting details.

The limitation of Markdown can be largely removed by Pandoc, but the problem is that Pandoc is a command-line tool. Power users may not find this to be a real problem, but the large number of command-line arguments can be overwhelming to beginners.

The goal of **rmarkdown** and R Markdown v2 is to provide quick conversion of R Markdown files into other document formats, using reasonably beautiful templates. The way that we achieve the goal is to wrap commonly used command-line arguments into R functions in **rmarkdown**. The main function in **rmarkdown** to render R Markdown documents to other document formats is *render()*. The first argument is the Rmd filename, and the second argument is the output format, which we will introduce in detail later in this chapter. For example, if you want to convert an R Markdown document foo.Rmd to Word, you only need to execute one line of code:

```
rmarkdown::render("foo.Rmd", "word_document")
```

You can certainly do it the hard way: first, call *knit()* in **knitr** to compile foo.Rmd to foo.md; then open a terminal or use the R function *system()* to execute a command like this, as we introduced in Section 13.2:

```
pandoc foo.md --output foo.docx \
  --from markdown+tex_math_single_backslash \
  --highlight-style tango
```

There are seven output format functions in **rmarkdown** at the moment: PDF, HTML, Word, Markdown, ioslides, Slidy, and Beamer. The first four are *document* formats, and the latter three are *presentation* formats. They are wrapper functions for both **knitr** and Pandoc, so you do not need to remember a lot of **knitr** options and Pandoc arguments — **knitr** chunk options and Pandoc command-line arguments are converted to **rmarkdown** function arguments. For example, the Pandoc argument --toc or --table-of-contents corresponds to the function argument toc = TRUE in **rmarkdown**.

In addition, **rmarkdown** has provided its own templates that aim to be visually pleasing by default. For example, for HTML output, it uses the Twitter Bootstrap styles and themes. Syntax highlighting for program code is also enabled by default.

The **rmarkdown** package is well supported in the RStudio IDE: you do not need to manually call the *render()* function, and you only need to click the Knit button on the toolbar. You can also set the output format and its options from a little GUI popped up through the gear button on the toolbar. If you wish to run **rmarkdown** outside of RStudio, you will want to learn more details about how **rmarkdown** works later.

Note RStudio has embedded Pandoc in it, so you do not need to install Pandoc separately if you use RStudio, otherwise you need to install Pandoc by yourself. If you have a separate installation of Pandoc, RStudio will use it only if your version is higher than RStudio's Pandoc version.

## 14.2 Pandoc's Markdown Extensions

First we introduce the syntax of Pandoc's Markdown. If you are familiar with R Markdown v1, you can still use its syntax with Pandoc, and the only significant change is how to write superscripts that are not math elements. In v1, you use a single caret, e.g., x^2. In Pandoc's Markdown, you need to surround the superscript with ^, e.g. x^2^. For math expressions, you still use one caret, e.g., $x^2$.

### 14.2.1 Basic Syntax

The syntax for other elements remains more or less the same in Pandoc's Markdown. For example, you use one # sign to write the first level section header, and two # signs for the second level header. Please review Section 5.2.1 for the syntax of basic elements in Markdown. Below are some new elements that may be useful (see http://johnmacfarlane.net/pandoc/ for the full documentation), and we show short examples of these elements under the bullets:

- Definition lists and example lists

```
A Special Term

: Describe/explain the term here.

(@) This is a numbered example.
(@) Another numbered example.

(@cool-example) This example is labeled.

This is a normal paragraph, and we can reference
the example (@cool-example) here.
```

- Footnotes using ^[...] and citations using [@id]

```
We write a nice description of X here^[Not to be
confused with Y], and X is useful.

Actually you should read the reference [@joe2014]
to know more about X. Here `joe2014` is a key in
the bibliography database.
```

- Figure/table captions

```
Pandoc has a Markdown extension named implicit_figures,
which is enabled by default. An image

![A figure caption.](path/to/image.png)

will be rendered to something like this in LaTeX:

\begin{figure}
  \includegraphics{path/to/image.png}
  \caption{A figure caption.}
\end{figure}

Similarly, you can add a table caption, e.g.

Table: This is a table caption.

--- ---- ----
A   B    C
--- ---- ----
a   10   bc
d   25   ef
--- ---- ----
```

- Raw TeX/HTML content

```
Sometimes you still feel Markdown is limited,
and you are so tempted to use LaTeX. That's
fine: you can write raw \TeX{} code in Markdown.

Markdown version:

![A long caption.](foo.png)
```

```
LaTeX version:

\begin{figure}
  \includegraphics[width=.8\textwidth]{foo.png}
  \caption[A short caption]{A long caption.}
\end{figure}

Pandoc can preserve the raw TeX content when
converting this document to LaTeX/PDF.
```

When using citations, you need to specify a bibliography database. If you are familiar with LATEX, you are likely to know BibTEX as well. The bibliography database can be a .bib file specified in the `bibliography` field in the YAML metadata (see next section). If you do not know BibTEX, you can embed the bibliography items in the YAML metadata using the `references` field (instead of `bibliography`), e.g.,

```
---
references:
  - id: joe2014
    title: A Nice Paper
    author:
      - family: Smith
        given: Joe
    issued:
      year: 2014
    container-title: The Journal of Awesome Research
    type: article-journal
  - id: john1980
    title: A Great Book
    author:
      - family: Brown
        given: John
    issued:
      year: 1980
    publisher: An Excellent Publisher
    type: book
---
```

Except for raw TEX/HTML code, all other elements are portable across all document formats. For example, a footnote `^[foo bar]` will be converted to `\footnote{foo}` when the output format is LATEX, and something like `<a href="#footnote-1"><sup>1</sup></a>` with the

link target `footnote-1` being a footnote item at the bottom of the page
when the output format is HTML. You should not expect raw TeX in
Markdown to be converted perfectly to Word, or raw HTML to be converted to Beamer, since raw TeX and HTML content can be fairly complicated, and perfect conversion is nearly impossible.

### 14.2.2  YAML Metadata

Another important extension in Pandoc's Markdown is the YAML metadata. YAML stands for "YAML Ain't Markup Language" or "Yet Another Markup Language," and it is basically a nested list structure. Pandoc uses YAML to write metadata of a document, such as the title, author, and date information. The metadata usually appears in the beginning of a document, and is enclosed between two lines of three dashes
`---`. Typical YAML metadata looks like this:

```
---
title: "A Nice Report"
author: "John Smith"
date: 2014/12/31
output:
  html_document:
    toc: yes
    number_sections: yes
  word_document: default
---

The body of the R Markdown document.
```

The most important field in the YAML metadata for **rmarkdown** is
the `output` field. This is where we specify the desired output format. If
it is missing, **rmarkdown** will assume the output format to be an HTML
document. If multiple formats are specified, the *render()* function will
use the first format by default, unless you have specified the second
argument of *render()* explicitly. You can also use `render('foo.Rmd',
'all')` to render all formats defined in the `output` field.

## 14.3  Output Formats

There is a series of format functions in **rmarkdown** with the suffixes
`_document` and `_presentation`, e.g., *html_document()*, *pdf_document()*,

and *beamer_presentation()*, etc. These functions can be used as the second argument of *render()*, e.g.,

```
library(rmarkdown)
render("foo.Rmd")
render("foo.Rmd", pdf_document())
render("foo.Rmd", word_document())
render("foo.Rmd", beamer_presentation())
render("foo.Rmd", ioslides_presentation())
```

Each output format function has its own arguments. For example, if you want to enable the table of contents for an HTML document, you can call:

```
library(rmarkdown)
render("foo.Rmd", html_document(toc = TRUE))
```

This is equivalent to providing the YAML metadata as:

```
---
output:
  html_document:
    toc: yes
---
```

In YAML, both yes and true mean the logical value TRUE. You can either use the YAML metadata and call *render()* without the second argument, or omit/ignore the YAML metadata and provide the second argument explicitly to *render()*. The YAML approach is more convenient and common; the output information is contained in the source document. The second approach can be useful when you want to override the output formats defined in YAML. See the help page of each output format function for what the possible options are, e.g., type ?rmarkdown::pdf_document in the R console to see the options for PDF output.

An output format function returns a list of options, including **knitr** package/chunk options, Pandoc arguments, and other auxiliary options for **rmarkdown**. We will explain them using *html_document()* as the example.

### 14.3.1 HTML Document

To see what *html_document()* really returns, you can run it and print the structure of the object returned:

```
library(rmarkdown)
str(html_document(), width = 55, strict.width = "wrap")

## List of 6
## $ knitr :List of 3
## ..$ opts_knit : NULL
## ..$ opts_chunk:List of 5
## .. ..$ dev : chr "png"
## .. ..$ dpi : num 96
## .. ..$ fig.width : num 7
## .. ..$ fig.height: num 5
## .. ..$ fig.retina: num 2
## ..$ knit_hooks: NULL
## $ pandoc :List of 5
## ..$ to : chr "html"
## ..$ from : chr
##    "markdown+autolink_bare_uris+ascii_identifiers+te"..
## ..$ args : chr [1:8] "--smart" "--email-obfuscation"
##    "none" "--self-contained" ...
## ..$ keep_tex: logi FALSE
## ..$ ext : NULL
## $ keep_md : logi FALSE
## $ clean_supporting: logi TRUE
## $ pre_processor :function (...)
## $ post_processor :function (metadata, input_file,
##    output_file, clean,
## verbose)
## - attr(*, "class")= chr "rmarkdown_output_format"
```

As you can see, *html_document()* has modified some of the **knitr** default chunk options, such as `fig.height` (**knitr**'s default is 7), and `fig.retina` (the original default is 1). These changes are for aesthetic reasons, although it is somewhat subjective to decide what kind of option values give better-looking results.

The list also contains Pandoc options: the output format is `html`, as you can see in the element `pandoc$to`; a few Pandoc arguments such as `--smart` and `--self-contained` are also included in the list.

There are some auxiliary options for **rmarkdown**, too. For example, `clean_supporting` means whether to clean up the intermediate output files after the HTML file has been rendered. Intermediate files may include figure files: if you want the HTML file to be self-contained, Pandoc will embed all external resources in it (such as images), so you no longer need these external files. In that case, *render()* will delete them after rendering the HTML file.

After we know the internals of an output format function, we can write our own format functions using different **knitr**/Pandoc options. We will introduce how to implement custom formats later in this chapter.

Now we show a full example of an R Markdown v2 document named Rmd-v2.Rmd. It is a little bit long, but it shows most of the features of Pandoc and **rmarkdown**.

```
---
title: "R Markdown v2 Demo"
author:
  - Li Lei
  - Han Meimei
date: "2015/01/01"
output:
  html_document:
    fig_caption: yes
  pdf_document:
    template: null
  word_document: default
bibliography: Rmd-v2.bib
---


# Start with a cool section

A bit _introduction_ here.

You can use traditional **Markdown** syntax, such as
[links](http://yihui.name/knitr) and `code`.

# Followed by another section

Of course you can write lists:

- apple
- pear
- banana

Or ordered lists:

1. items
```

```
1. will
1. be
1. ordered
    - nested
    - items
```

```
# More sections
```

```
## Hi
```

```
hi hi
```

```
## Hello
```

```
hello hello
```

```
## Howdy
```

```
howdy howdy
```

```
# Okay, some R code
```

````
```{r linear-model}
fit = lm(dist ~ speed, data = cars)
b = coef(fit)  # coefficients
summary(fit)
```
````

```
The code will be highlighted in all output formats.
```

```
# And some pictures
```

````
```{r lm-vis, fig.cap='Regression diagnostics'}
par(mfrow = c(2, 2), pch = 20, mar = c(4, 4, 2, .1),
    bg = 'white')
plot(fit)
```
````

```
# A little bit math
```

Our regression equation is $Y=`r b[1]`+`r b[2]`x$, and the
model is:

```
$$ Y = \beta_0 + \beta_1 x + \epsilon$$

# Pandoc extension: definition lists

Programmer
:   A programmer is the one who turns coffee into code.

LaTeX
:   A simple language with a couple of backslashes.

# Pandoc extension: examples

We have some examples.

(@) Think what is `0.3 + 0.4 - 0.7`. Zero. Easy.
(@weird) Now think what is `0.3 - 0.7 + 0.4`. Still zero?

People are often surprised by (@weird).

# Pandoc extension: tables

A table here.

Table:  Demonstration of simple table syntax.

```{r echo=FALSE}
knitr::kable(head(iris))
```

# Pandoc extension: footnotes

We can also write footnotes[^1].

[^1]: hi, I'm a footnote

Or write some inline footnotes^[as you can see here].

# Pandoc extension: citations

We compile the R Markdown file to Markdown through **knitr**
[@R-knitr] in R [@R-base]. For more about @R-knitr,
see <http://yihui.name/knitr>.
```

FIGURE 14.1: A preview of the HTML output document from R Markdown v2 in an RStudio window.

```
# References

```{r include=FALSE}
knitr::write_bib(c('base', 'knitr'), 'Rmd-v2.bib')
```
```

You may need to review the sections 6.3 and 12.4.1 if you are not sure about how *kable()* or *write_bib()* works.

Figure 14.1 is a preview of the HTML output document after we render this example in RStudio. It shows the title, author, date, and the first few sections of the document. That is the default Twitter Bootstrap style in **rmarkdown**. Figure 14.2 is a preview of the last few sections. Even though footnotes and citations are not native elements of HTML (they may be natural to LaTeX users), Pandoc managed to generate them in HTML anyway.

There is a large number of options that you can tweak for the HTML output. See the help page ?rmarkdown::html_document for a full list.

Demonstration of simple table syntax.

| Sepal.Length | Sepal.Width | Petal.Length | Petal.Width | Species |
|---:|---:|---:|---:|---|
| 5.1 | 3.5 | 1.4 | 0.2 | setosa |
| 4.9 | 3.0 | 1.4 | 0.2 | setosa |
| 4.7 | 3.2 | 1.3 | 0.2 | setosa |
| 4.6 | 3.1 | 1.5 | 0.2 | setosa |
| 5.0 | 3.6 | 1.4 | 0.2 | setosa |
| 5.4 | 3.9 | 1.7 | 0.4 | setosa |

# Pandoc extension: footnotes

We can also write footnotes[1].

Or write some inline footnotes[2].

# Pandoc extension: citations

We compile the R Markdown file to Markdown through **knitr** (Xie 2014) in R (R Core Team 2014). For more about Xie (2014), see http://yihui.name/knitr.

# References

R Core Team. 2014. *R: A Language and Environment for Statistical Computing*. Vienna, Austria: R Foundation for Statistical Computing. http://www.R-project.org/.

Xie, Yihui. 2014. *Knitr: A General-Purpose Package for Dynamic Report Generation in R*. http://yihui.name/knitr/.

1. hi, I'm a footnote↵
2. as you can see here↵

FIGURE 14.2: A preview of the table, footnotes, and citations: the table was generated by *kable()*, and the bibliography database was created from *write_bib()* in **knitr**.

For example, we change the CSS theme using the `theme` field, add a table of contents using the `toc` field, and number the section titles using the `number_sections` field in YAML (Figure 14.3):

```
---
output:
  html_document:
    fig_caption: yes
    number_sections: yes
    theme: readable
    toc: yes
---
```

Currently these CSS themes are available in **rmarkdown** (you can see a preview at http://bootswatch.com):

```
## [1] "default"  "cerulean" "journal"  "flatly"
## [5] "readable" "spacelab" "united"   "cosmo"
```

If you need to further tweak the appearance of the output, you can apply your own CSS files using the `css` field, e.g.,

```
---
output:
  html_document:
    css: my_own.css
---
```

If you just want to use your own CSS and do not want any themes (including syntax highlighting themes) from **rmarkdown**, you can remove them completely by specifying `theme` and `highlight` to be `null`:

```
---
output:
  html_document:
    css: my_own.css
    theme: null
    highlight: null
---
```

Because an HTML page often has external dependencies, such as CSS, JavaScript, and image files, it may be inconvenient when you share the HTML file with other people, because you have to make sure these dependencies are also included when you send the HTML file to them.

# R Markdown v2 Demo

## Li Lei

## Han Meimei

## 2015/01/01

- 1 Start with a cool section
- 2 Followed by another section
- 3 More sections
  - 3.1 Hi
  - 3.2 Hello
  - 3.3 Howdy
- 4 Okay, some R code
- 5 And some pictures
- 6 A little bit math
- 7 Pandoc extension: definition lists
- 8 Pandoc extension: examples
- 9 Pandoc extension: tables
- 10 Pandoc extension: footnotes
- 11 Pandoc extension: citations

# 1 Start with a cool section

A bit *introduction* here.

You can use traditional **Markdown** syntax, such as links and code .

# 2 Followed by another section

Of course you can write lists:

FIGURE 14.3: A preview of the "readable" theme (you can see the fonts are different with Figure 14.1), with a table of contents and numbered sections.

Pandoc has an option to make the HTML file self-contained by embedding all external dependencies into the HTML file. For example, JavaScript files are read into the HTML file, and images are base64 encoded. You can share a self-contained HTML file just like a PDF file; everything you need has been embedded into a single file. In **rmarkdown**, this is controlled by the option self_contained. When you have multiple Rmd files to be rendered by **rmarkdown**, it may be a good idea to turn off the self-contained mode, otherwise there will be a lot of redundancy since some external dependencies may be embedded into every single HTML output file. When the self-contained mode is off, you can put the shared dependencies into a common directory, specified via the lib_dir option, e.g.,

```
---
output:
  html_document:
    self_contained: no
    lib_dir: assets
---
```

Sometimes you may want to include additional content in the HTML header, before the body, or after the body of the document. In these cases, **rmarkdown** has an option includes in which you can specify the filenames of the additional content. Suppose you want to use the JavaScript library D3 (http://d3js.org) in the HTML output, then you can write this in a file doc_header.html:

```
<script src="http://d3js.org/d3.v3.min.js" charset="utf-8">
</script>
```

You also have two files doc_before.html and doc_after.html, which are the content to be inserted before and after the body, respectively. For example, you may want to write a navigation menu in doc_before.html, and some copyright information in doc_after.html. These three files can be included in the HTML output file by:

```
---
output:
  html_document:
    includes:
      in_header: doc_header.html
      before_body: doc_before.html
      after_body: doc_after.html
---
```

For any output format, Pandoc needs a template to create the output file. There are several Pandoc variables available in the template, and you can use these variables to define your own template. For example, this can be a minimal HTML template:

```
<html>

  <head>
    <title>$title$</title>
  </head>

  <body>
  $body$
  </body>

</html>
```

We only used two variables $title$ and $body$ in this template. The first variable contains the document title specified in the `title` field in the YAML metadata. The second variable is the body of the Markdown document after it is converted to HTML. You can learn more possible variables from either the **rmarkdown** source package (`https://github.com/rstudio/rmarkdown`) or Pandoc's default templates (`https://github.com/jgm/pandoc-templates`).

To use a custom template, you can use the `template` field in YAML, e.g.,

```
---
output:
  html_document:
    template: my_template.html
---
```

Finally, you can customize command-line arguments to be passed to Pandoc in the `pandoc_args` field. As a matter of fact, the R arguments in *html_document()* are eventually converted to Pandoc arguments. For example, the R argument `self_contained = TRUE` (or `self_contained: yes` in YAML) is equivalent to the Pandoc argument `--self-contained`, and also equivalent to this in YAML:

```
---
output:
  html_document:
    pandoc_args: "--self-contained"
---
```

So far we have covered most of the possibilities to customize the output on the Pandoc's Markdown side. It is also possible to customize **knitr** chunk options in YAML. Currently there are four chunk options that you can set in YAML:

**fig_width, fig_height** the default size of the figures

**fig_retina** a scaling ratio for Retina displays; the default is 2 in **rmarkdown**, which means a figure of the size $m \times n$ has an actual size of $2m \times 2n$, but is scaled to half of its actual size in the output (this can improve the image qualities on Retina displays)

**fig_caption** whether to render and show figure captions (this basically means the `figure` environment with `\caption{}` when the output format is LaTeX); if `FALSE`, you will not see the figure caption in HTML output, since the caption will be put in the `alt` attribute of the `<img>` tag, which is invisible

Apparently, the `fig_retina` option will make the file size of images larger in return for the image quality. You can try `fig_retina = TRUE` and `FALSE` separately, and see if you can notice any differences on your device.

### 14.3.2   LaTeX/PDF Document

Once you are familiar with the HTML document format, it will be easy for you to master other output formats, because many options are common in these formats. For example, you can also use the options such as `fig_width`, `fig_height`, `toc`, `number_sections`, and `highlight` in *pdf_document()*. In this section, we only focus on the options that are specific to PDF document output.

Figure 14.4 is a preview of a page in the PDF output from the same example we used in the previous section. It does not look too much different from Figure 14.2. For the same R Markdown document, everything that worked in the HTML output still works in LaTeX/PDF, including section headings, tables, footnotes, and citations, etc.

Similarly, we can add a table of contents, and number the sections as we did for the HTML output (Figure 14.5):

```
---
output:
  pdf_document:
    number_sections: yes
    toc: yes
---
```

## Pandoc extension: tables

A table here.

Table 1: Demonstration of simple table syntax.

| Sepal.Length | Sepal.Width | Petal.Length | Petal.Width | Species |
|---:|---:|---:|---:|---|
| 5.1 | 3.5 | 1.4 | 0.2 | setosa |
| 4.9 | 3.0 | 1.4 | 0.2 | setosa |
| 4.7 | 3.2 | 1.3 | 0.2 | setosa |
| 4.6 | 3.1 | 1.5 | 0.2 | setosa |
| 5.0 | 3.6 | 1.4 | 0.2 | setosa |
| 5.4 | 3.9 | 1.7 | 0.4 | setosa |

## Pandoc extension: footnotes

We can also write footnotes[1].

Or write some inline footnotes[2].

## Pandoc extension: citations

We compile the R Markdown file to Markdown through **knitr** (Xie 2014) in R (R Core Team 2014). For more about Xie (2014), see http://yihui.name/knitr.

## References

R Core Team. 2014. *R: A Language and Environment for Statistical Computing.* Vienna, Austria: R Foundation for Statistical Computing. http://www.R-project. org/.

Xie, Yihui. 2014. *Knitr: A General-Purpose Package for Dynamic Report Generation in R.* http://yihui.name/knitr/.

---

[1] hi, I'm a footnote
[2] as you can see here

FIGURE 14.4: A preview of the 4th page of the PDF output document from the R Markdown v2 example.

# R Markdown v2 Demo

Li Lei       Han Meimei

2015/01/01

## Contents

FIGURE 14.5: A preview of the PDF output document, with a table of contents and numbered sections.

Pandoc has a few LaTeX-specific options that you can use in the YAML metadata, and you can find the full documentation on the Pandoc website. We only list a few of them here:

**fontsize** the font size of the document, e.g., 10pt, 11pt, 12pt

**documentclass** the document class, e.g., article, book, report

**classoption** options for the document class, e.g., a4paper, twocolumn

**geometry** options for the **geometry** package, e.g., tmargin=2cm, bmargin=2cm, lmargin=3cm, rmargin=3cm

Note these are top-level options in YAML, and you should not put them under the `pdf_document` field.

The default LaTeX engine is `pdflatex`, and you can change it via the `latex_engine` option in *pdf_document()*. Currently possible engines are `pdflatex`, `xelatex`, and `lualatex`. You may also preserve the intermediate LaTeX output file via the `keep_tex` option, which can be useful for debugging and other purposes.

Below is an example of the YAML metadata for a document that uses the `book` class, a font size of 11pt, a two-column layout, custom margin settings, the XeLaTeX engine, and also preserves the LaTeX file:

```
---
documentclass: book
classoption: twocolumn
fontsize: 11pt
geometry:
  - tmargin=2cm
  - bmargin=2cm
  - lmargin=3cm
  - rmargin=3cm
output:
  pdf_document:
    latex_engine: xelatex
    keep_tex: yes
---
```

We have introduced the `includes` and `template` options in the previous section, and they may be more useful for LaTeX output, because it is very common for LaTeX users to customize the output using certain LaTeX packages in the preamble. You can put such content in an external file, and include it in the preamble via the `in_header` option under the `includes` option. If you are not satisfied with the default

LATEX template, you can just write your own. Before you really do it, please check the Pandoc documentation carefully to see if you can get what you want by YAML options. It is relatively easy to write a new LATEX template, but it may not be trivial to maintain it in the future, since you need to be aware of possible future changes in Pandoc.

### 14.3.3   Word Document

There are not many options to customize for Word documents. You can still set the figure size, and syntax highlighting themes, etc. Figure 14.6 shows the Word output from the example in Microsoft Word 2013.

The most important and useful feature for Word documents is perhaps the template. For other document formats, you can provide a plain text template, but you cannot easily do so for Word, because a Word document is a relatively complicated binary file. However, Pandoc allows you to provide a Word document as its "reference document," which is essentially a style template. This reference document must be based on one of Pandoc's Word output documents, in which you update its styles for different elements. Note only the styles defined in the document will be used, and the content will be largely ignored.

We have prepared a short video at `https://vimeo.com/110804387` to show you how to define styles in Word documents. You can also see Figure 14.7 and 14.8. The basic steps are:

1. Create an arbitrary Word document using Pandoc, e.g., use `word_document` as the `output` option in the YAML metadata;

2. Open the Word document, and find the "Styles" panel indicated in Figure 14.7;

3. Put the cursor on the element of which you want to modify the style, and there should be an item in the Styles panel highlighted;

4. Open the item by clicking the ¶ symbol on the right, and you will see a window like Figure 14.8. That is where you can modify the styles. For example, you can change the font family of the title element to be Bookman Old Style.

After you update the styles of this Word document, you can save it (say, as template.docx under the same directory as the Rmd file) and use it as the reference document:

```
---
output:
  word_document:
```

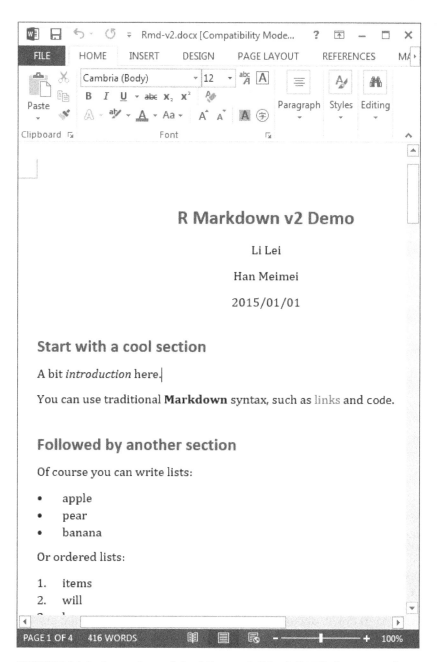

FIGURE 14.6: A preview of the Microsoft Word (2013) document from R Markdown v2.

FIGURE 14.7: Open the styles panel in Word: find a pane named "Styles" on the toolbar, and expand it to a floating panel.

```
    renference_docx: template.docx
---
```

Besides the styles of the elements, the styles of the layout can also be respected if you use Pandoc >= 1.13. For example, the margins, page size, page orientation, header, and footer in the reference document will be carried over to the new Word document.

### 14.3.4   Markdown Documents

An R Markdown document can be converted to different flavors of Markdown documents, such as Pandoc's Markdown, the original (strict) Markdown, Github Flavored Markdown, MultiMarkdown, and PHP Markdown Extra. You can use the function *md_document()* for *render()* or output: md_document in YAML. The main option for md_document is variant, which specified which flavor of Markdown you want.

FIGURE 14.8: Modify styles of elements in Word: you can change the font family, font size, font style, and color, etc.

### 14.3.5 ioslides Presentation

R Markdown can be used to create slides for presentation purposes. With the process of Web technologies, HTML5 slides seem to be popular nowadays. You can present slides in a Web browser. This is convenient since you do not need special software packages to display the slides, and you can find a Web browser almost everywhere. This is not true for proprietary software such as Microsoft PowerPoint or Keynote for Mac.

There are two types of built-in HTML5 presentation formats in **rmarkdown**: ioslides and Slidy. You can extend **rmarkdown** to use your own favorite HTML5 presentation library.

For ioslides, each first-level section heading will create a separate

FIGURE 14.9: The title slide of an ioslides presentation: you can also use the table of contents in RStudio to navigate through the slides.

slide with a dark background by default; each second-level heading creates a new slide with the content of this section on it. If you do not want a section heading, you can create a new slide with three dashes ---. Figure 14.9 is a screenshot of ioslides in the RStudio preview window, created using the same example as previous sections and the YAML metadata (if you really try this example, you may want to remove the content between the first-level heading and second-level heading):

```
---
output:
  ioslides_presentation: default
---
```

When you do the presentation, you may want to use the fullscreen mode, which can be turned on by the keyboard shortcut f (just press the F key). The key W toggles the widescreen mode. If the slide size is too big or too small, you can zoom in/out the page. Normally you can do it by holding the Ctrl (or Command) key, then press Plus (+) or Minus (-).

There are a few options for the `ioslides_presentation` format you can use to tweak the appearance of the slides:

**incremental** (yes/no) whether to show bullets incrementally

**logo** an image that you want to use as the logo in the slides (it will be displayed in the footer of each slide)

**css** a custom CSS file

You can also customize each slide individually. For example, if you put a token `{.build}` after a second-level section heading, the elements on this page will be displayed incrementally as you proceed in the presentation, e.g.,

```
## A new slide {.build}

First show this.

Then show that.

Finally show a funny GIF animation.

![](foo.png)
```

HTML5 slides are usually for presentation instead of printing purposes. However, you may also print the slides as PDFs from your Web browser. At the moment, we recommend you to use Google Chrome if you want to print the slides. You should expect the appearance of printed slides to differ from that of the displayed slides.

### 14.3.6 Slidy Presentation

The rules of writing slides for Slidy are the same as ioslides. The function for Slidy presentation output in **rmarkdown** is *slidy_presentation()*. Figure 14.10 shows one slide of the Slidy presentation created from the R Markdown example.

A few keyboard shortcuts are available, e.g., press C to see the table of contents, S to make the font smaller, and B to make the font bigger, etc.

FIGURE 14.10: One slide from the Slidy presentation generated from the R Markdown example: you can also click "Contents" at the bottom to show the table of contents.

Besides the `incremental` and `css` options we mentioned before, Slidy has some additional features that may be useful, including the options:

**duration** sets a countdown timer in the footer to remind you of the time, e.g., if you have a 50-minute talk, you can set `duration:   50` in YAML

**footer** a custom message in the footer, e.g., you can display the name of your institute or copyright information

To print Slidy slides, you can also use Google Chrome.

### 14.3.7   Beamer Presentation

Beamer, introduced in Section 12.3.4 is a LaTeX application, so you can build an Rnw file as a LaTeX document with code chunks shown in Section 12.3.4 and compile directly into the PDF format. Markdown is simpler and faster for all but veteran LaTeX users, so we recommend trying it with the `beamer_presentation` format. If you need some of the more advanced Beamer or LaTeX features, they can be added within Markdown as Pandoc supports LaTeX code within Markdown.

Figure 14.11 shows two slides of the Beamer presentation created from the previous R Markdown example. All we did was change the YAML metadata to:

```
---
title: "R Markdown v2 Demo"
author:
  - Li Lei
  - Han Meimei
date: "2015/01/01"
output:
  beamer_presentation:
    theme: AnnArbor
bibliography: Rmd-v2.bib
---
```

If we were to write the slides in raw LaTeX, the source document would be like this:

```
\documentclass{beamer}
\usetheme{AnnArbor}

\title{R Markdown v2 Demo}
\author{Li Lei \and Han Meimei}
\date{2015/01/01}

\begin{document}
\frame{\titlepage}

\begin{frame}{Start with a cool section}

A bit \emph{introduction} here.

You can use traditional \textbf{Markdown} syntax, such as
\href{http://yihui.name/knitr}{links} and \texttt{code}.

\end{frame}

\begin{frame}{Followed by another section}
```

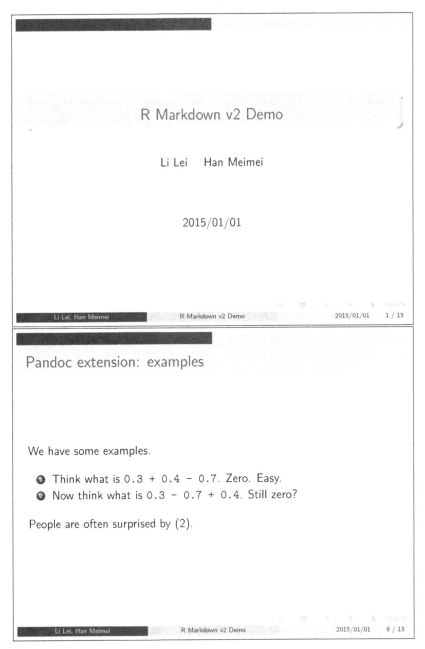

FIGURE 14.11: Two slides from the Beamer presentation created by R Markdown: the title slide, and the slide that shows the Pandoc extension of the example environment.

```
Of course you can write lists:

\begin{itemize}
\item
  apple
\item
  pear
\item
  banana
\end{itemize}

....

\end{document}
```

Compare that with the R Markdown source code in Section 14.3.1, and hopefully you see how much more code you would have to type when writing in raw LaTeX than writing in Markdown.

Each new slide is a new section in Markdown, and the level of the section is determined by the highest level in the document hierarchy that is followed immediately by the slide content. In the following example, each first-level section (#) is a new slide:

```
---
output: beamer_presentation
---

# One Section

- content
- content

# Another Section

![](foo.png)
```

And in this example, each sub-section (##) is a new slide:

```
---
output: beamer_presentation
---

# One Section
```

```
## One Sub-section

- content
- content

# Another Section

## Another Sub-section

![](foo.png)
```

To display list items incrementally, you can use the `incremental` option just like what we can do for ioslides and Slidy presentations. Other options such as `toc`, `highlight`, `fig_width`, `fig_height`, `fig_caption`, `includes`, and `template` have been explained in previous sections.

There are many themes (including font themes and color themes) in Beamer. You can use them via the `theme`, `fonttheme`, and `colortheme` options. Figure 14.11 used the `AnnArbor` theme, and default font/color themes. If you use RStudio, you can choose these themes from the GUI, so you do not need to remember the many theme names.

### 14.3.8  Other Formats

Besides the document and presentation formats, **rmarkdown** also has two special output formats: *html_vignette()* for HTML package vignettes (Section 15.4) and *tufte_handout()* for the Tufte handout (here Tufte refers to Edward R. Tufte).

The *html_vignette()* format is a wrapper of *html_document()*, with a special CSS theme; the file size of the HTML vignette produced by *html_document()* is too big because it contains the Twitter Bootstrap assets, the jQuery library, and highlight.js by default. The *html_vignette()* format has removed all these components, and uses a single lightweight CSS file. The option `fig_retina` has been set to 1 to further reduce the image file sizes. This format function is a good example of how to build your own format based on existing format functions, and its source code is very simple:

```
html_vignette <- function(fig_width = 3,
    fig_height = 3, dev = "png", css = NULL,
    ...) {
    if (is.null(css)) {
        css <- system.file("rmarkdown", "templates",
```

```
              "html_vignette", "resources",
              "vignette.css", package = "rmarkdown")
  }
  html_document(fig_width = fig_width,
      fig_height = fig_height, dev = dev,
      fig_retina = FALSE, css = css, theme = NULL,
      highlight = "pygments", ...)
}
```

The *tufte_handout()* format is a wrapper for the LaTeX document class tufte-handout.cls. The most notable characteristics of the Tufte handout style are perhaps the use of sidenotes, and the well-designed typography. See Figure 14.12 for an example page. Its YAML metadata is this:

```
---
title: "Tufte Handout"
author: "John Smith"
date: "August 13th, 2014"
output: rmarkdown::tufte_handout
---
```

## 14.4  Interactive Documents with Shiny

Shiny (Chang et al., 2015) is a Web application framework that makes it easy to create interactive apps using R. You can create a Web user interface (UI) using Shiny UI functions, e.g., text input boxes, drop-down lists, radio buttons, and sliders, etc. These UI elements can interact with R after you specify the server logic in R, e.g., after you click a button, what you expect R to do. If you are not familiar with Shiny, please check out the website http://shiny.rstudio.com to learn the basics about Shiny.

Because a Shiny app is basically an HTML page, and it happens that R Markdown can be rendered to HTML, too, it is possible to combine R Markdown and Shiny in one document. We call such documents "interactive documents," since they contain interactive components from Shiny. Figure 14.13 shows a minimal example of an interactive document. Its source document is as follows:

*Tufte Handout*
*John Smith*
*August 13th, 2014*

*Introduction*

The Tufte-LaTeX [1] document classes define a style similar to the style Edward Tufte uses in his books and handouts. Tufte's style is known for its extensive use of sidenotes, tight integration of graphics with text, and well-set typography.

[1] https://code.google.com/p/tufte-latex/

*Headings*

This style provides a- and b-heads (that is, # and ##), demonstrated above. An error is emitted if you try to use ### and smaller headings.

IN HIS LATER BOOKS[2], Tufte starts each section with a bit of vertical space, a non-indented paragraph, and sets the first few words of the sentence in small caps. To accomplish this using this style, use the \newthought command as demonstrated at the beginning of this paragraph.

[2] http://www.edwardtufte.com/tufte/books_be

*Figures*

*Margin Figures*

Images and graphics play an integral role in Tufte's work. To place figures or tables in the margin you can use the fig.margin knitr chunk option. For example:

```
library(ggplot2)
qplot(Sepal.Length, Petal.Length, data = iris,
    color = Species)
```

Note the use of the fig.cap chunk option to provide a figure caption. You can adjust the proportions of figures using the fig.width and fig.height chunk options. These are specified in inches, and will be automatically scaled down to fit within the handout margin.

Figure 1: Sepal length vs. petal length, colored by species

*Equations*

You can also include LaTeX equations in the margin by explicitly invoking the marginfigure environment.

Note the use of the \caption command to add additional text below the equation.

$$\frac{d}{dx}\left(\int_0^x f(u)\,du\right) = f(x).$$

Figure 2: An equation

FIGURE 14.12: An example page using the Tufte handout style: you can arrange elements into the side margin, such as footnotes, figures, equations, and so on.

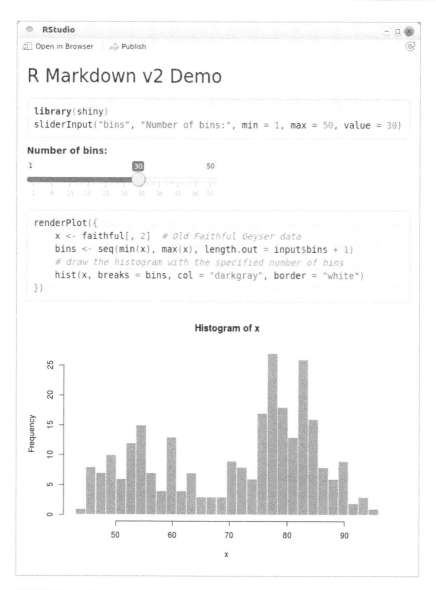

FIGURE 14.13: A simple interactive document using R Markdown and Shiny: you can change the value of the slider, and the number of bins in the histogram will be automatically changed.

```
---
title: "R Markdown v2 Demo"
runtime: shiny
output: html_document
---

```{r}
library(shiny)
sliderInput("bins", "Number of bins:", min = 1, max = 50,
            value = 30)

renderPlot({
  x <- faithful[, 2]  # Old Faithful Geyser data
  bins <- seq(min(x), max(x), length.out = input$bins + 1)
  # draw the histogram with the specified number of bins
  hist(x, breaks = bins, col = 'darkgray', border = 'white')
})
```
```

To turn a normal R Markdown document into an interactive document, you only need to add the option `runtime:` shiny in the YAML metadata. Then you can use functions in the **shiny** package. In the above example, we created a slider on the HTML page using *sliderInput()*, which is a UI function in **shiny**. The id of the slider is `bins`. Then we rendered a histogram using the *renderPlot()* function. The most important bit in this code chunk is `input$bins`, which is a variable value associated with the slider with the id `bins`. When we update the value of the slider, its value will be passed to the expression in *renderPlot()*, and the plot will be redrawn accordingly.

Instead of *render()*, interactive documents should be compiled by the *run()* function in **rmarkdown**. If you use RStudio, you will see that the label of Knit button on the toolbar becomes Run Document after you add `runtime:` shiny to an R Markdown document, and you can click the button to run the document.

Not all Shiny apps can be so simple as the one in Figure 14.13. When you have several UI elements, you may want to arrange them in a separate app instead of writing them out in code chunks linearly. The function *shinyApp()* in **shiny** allows you to build a full app by specifying all UI elements and the server logic in one function. Then you can either embed full apps using *shinyApp()* explicitly in R Markdown, or write your own function that returns a *shinyApp()* object, so that other people can easily use your app as well.

Static HTML documents can be uploaded to any website or emailed when you want to share them. For interactive documents, there must be an active R session running behind them. One possible way to share interactive documents is to publish them to `http://shinyapps.io`, which is hosted by RStudio. If you do not want to publish to this website, you can set up your own Shiny Server: `http://www.rstudio.com/products/shiny/shiny-server/`.

## 14.5   Extending R Markdown v2

If none of the output format functions meet your need, you can extend them or write a completely new format. Before you do it, please make sure you have looked at all the possibilities in the existing output formats. Sometimes there is no need to invent anything new. For example, if all you want is to use a different LaTeX document class, you may as well set the `documentclass` option in the YAML metadata, although you can certainly also write a new template with the desired document class. Take the Tufte handout as an example:

```
---
title: "R Markdown v2 Demo"
author: John Smith
date: "2015/01/01"
output: pdf_document
documentclass: tufte-handout
classoption: nohyper
geometry: no
---
```

The above YAML metadata makes use of the existing *pdf_document()* format. Alternatively, you can prepare a template like:

```
\documentclass{tufte-handout}

$if(title)$
\title{$title$}
$endif$
$if(author)$
\author{$for(author)$$author$$sep$ \and $endfor$}
$endif$
$if(date)$
\date{$date$}
```

```
$endif$

\begin{document}

$if(title)$
\maketitle
$endif$

$body$

\end{document}
```

Then use the `template` option in `pdf_document`. There are a number of disadvantages of writing a custom template like that:

- Pandoc's default LaTeX is much more flexible (`https://github.com/jgm/pandoc-templates`), which can also deal with the table of contents, the list of figures, and the abstract, etc.;

- It requires more work to write a new template than to use existing options in YAML;

- After you write a template, you will have to watch out for future changes in Pandoc, which may break your template, or you may miss some useful new features. By comparison, if you use Pandoc's templates, you do not need to maintain them.

Then you may ask why we have the *tufte_handout()* format in **rmarkdown** after all. Actually what this new format does is more than just a LaTeX template: it also defines a few **knitr** chunk options to produce full-width figures (`fig.fullwidth` = TRUE) and margin figures (`fig.margin` = TRUE). Existing output formats do not provide these two different figure types.

### 14.5.1   Templates

The first type of **rmarkdown** extension is to define a new template. We have shown an example above for the Tufte handout, and also an example earlier in Section 14.3.1 for HTML document output.

The repository `https://github.com/jgm/pandoc-templates` contains all templates used by Pandoc, and you can also take a look at the custom templates in the **rmarkdown** source package at `https://github.com/rstudio/rmarkdown`. If there are any template variables that you do not understand, you can check out the documentation at `http://johnmacfarlane.net/pandoc/`.

To share a template with other users, the easiest way is to put it in an R package under the inst/rmarkdown/templates/ directory. You can create a new directory, say, my_template, and put the template file under it. Your template may require certain dependencies, such as CSS/JavaScript files, or LaTeX packages. They can be collected under a sub-directory skeleton/ under my_template. In the skeleton/ directory, you can also provide a sample Rmd file skeleton.Rmd. Finally, you can describe the template in a YAML file template.yaml under my_template with three YAML fields:

**name** the name of the template, e.g., *"Journal of Statistical Software"*;

**description** a short description of the template, e.g., "This is a template for JSS articles";

**create_dir** yes or no, or true or false (to be explained soon);

Suppose you installed such an R package named **myPackage**, then you can create a new draft from the template using the *draft()* function:

```
rmarkdown::draft("my_article.Rmd", template = "my_template",
    package = "myPackage")
```

This function looks for the template my_template in **myPackage**, copies skeleton.Rmd as my_article.Rmd to the current working directory, and also copies the dependencies. The YAML option create_dir mentioned above determines whether to create a new directory for the draft my_article.Rmd.

RStudio has made this process even easier. From the menu File ▷ New File ▷ R Markdown, you can see all templates in all locally installed packages (Figure 14.14).

The **rticles** package (https://github.com/rstudio/rticles) is a collection of templates for several LaTeX document classes. You can use its templates to write papers in R Markdown for the *Journal of Statistical Software*, and *The R Journal*, etc.

## 14.5.2 New Formats

The second type of **rmarkdown** extension is new output formats. The new format can be based on an existing output format, or a completely new format. The former is easy: you just define an R function that returns an output format object, with certain options modified from an existing output format function. As a minimal example, we create a function html_toc below, turning the default value of the toc argument from FALSE to TRUE:

FIGURE 14.14: Create a new R Markdown document from templates: you can select a template from the list.

```
html_toc <- function(toc = TRUE, ...) {
    rmarkdown::html_document(toc = toc, ...)
}
```

A new format function should be put in an R package (we still assume its name is **myPackage**), and then you can use it in YAML. Here are two examples:

```
---
output: myPackage::html_toc
---
```

```
---
output:
  myPackage::html_toc:
    toc: no
    self_contained: no
---
```

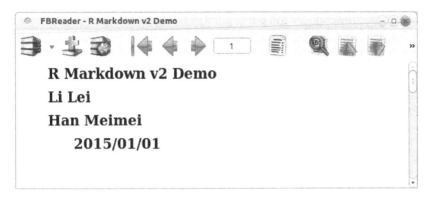

FIGURE 14.15: Create an E-book from R Markdown: this figure shows the title page of the EPUB book in FBReader (a free E-book reader).

For the second example, what will be called when we render this Rmd file is:

```
rmarkdown::render("foo.Rmd", myPackage::html_doc(toc = FALSE,
    self_contained = FALSE))
# which is essentially render('foo.Rmd',
# html_document(toc = FALSE, self_contained = FALSE))
```

As we explained in Section 14.3.1, the output format is a list of three types of options: **knitr** options, Pandoc options, and **rmarkdown** options. We customized the Pandoc toc in the above minimal example, and you can certainly customize more options in the output format function. There are a few helper functions *output_format()*, *knitr_options()*, and *pandoc_options()* in **rmarkdown** that you can use to compose the output format. See the repository https://github.com/jjallaire/revealjs for an example of how to create a new format for reveal.js (an HTML5 presentation format). Below we show a minimal example of how to create an output for EPUB (an E-book format):

```
#' @importFrom rmarkdown output_format
#' @importFrom rmarkdown knitr_options
#' @importFrom rmarkdown pandoc_options
epub_book <- function(to = c("epub", "epub3")) {
    to <- match.arg(to)
    optk <- knitr_options()
    optp <- pandoc_options(to, ext = ".epub")
    output_format(knitr = optk, pandoc = optp)
}
```

Put this function in the package **myPackage**, and you will be able to create E-books from R Markdown. Here is a minimal R Markdown example (Figure 14.15):

```
---
title: "R Markdown v2 Demo"
author:
  - Li Lei
  - Han Meimei
date: "2015/01/01"
output: myPackage::epub_book
---


# Start with a cool section

```{r}
1 + 1
```
```

The key in the format function *epub_book()* was to specify the argument to of *pandoc_options()* to be either epub or epub3. Pandoc supports a large number of document formats, and **rmarkdown** only included a small subset of them. You can build your own format function using the approach introduced above.

### 14.5.3   HTML Widgets

We explained the includes option in the YAML metadata in Section 14.3.1. When you want to include JavaScript libraries in the HTML document output, you can use the includes option. There are two disadvantages of this approach:

1.  It is not portable, in the sense that when you share the R Markdown document with other people, you should remember to copy the dependencies specified in the includes option; it is not convenient for other people to reuse your dependencies, either;

2.  You have to write (sometimes a lot of) JavaScript code in R Markdown to call the JavaScript libraries, but not all R users are familiar with JavaScript, so they may not be able to work on the R Markdown document.

The idea of HTML widgets is to provide native R interfaces to JavaScript libraries, so that even those who do not understand JavaScript can still use the libraries without worrying about the underlying dependencies or JavaScript syntax. When you draw a plot using a JavaScript library, all you need to do is call an R function in a code chunk.

The **htmlwidgets** package (Vaidyanathan et al., 2014) was designed for package developers to port JavaScript libraries into R easily. It is well-documented at `http://www.htmlwidgets.org`, and you can see several example packages on the website, too. We will not describe the technical details here, and we just show a quick example of what an HTML widget looks like. Here is a minimal R Markdown example (you need to install the **DT** package from `https://github.com/rstudio/DT` before trying this example):

```
---
title: "R Markdown v2 Demo"
author:
  - Li Lei
  - Han Meimei
date: "2015/01/01"
output: html_document
---


Here is a table generated by the DataTables library.

```{r}
DT::datatable(iris)
```
```

Figure 14.16 shows the output. The **DT** package is an interface to the JavaScript library DataTables (`http://datatables.net`). As you can see, the R Markdown source document is really simple, and you do not see the JavaScript files or any JavaScript code at all. You simply call the function *datatable()*, and your data frame will be displayed via DataTables. The hard work of passing data to the HTML page, parsing and rendering it has been done by the package authors, and users do not have to understand all the underlying technical details.

## 14.6   Changes in R Markdown from v1 to v2

If you happen to have started using R Markdown when it was v1, here is a list of changes that you should be aware of when you transition from v1 to v2:

FIGURE 14.16: A table created by the DataTables library in R Markdown: you can order the columns, search in the table, and the full table can be displayed on multiple pages.

- The **knitr** package is no longer loaded (strictly speaking, *attached*) by default in v2, which means the functions and objects in the **knitr** package are not available unless you explicitly load the package, e.g., via the command `library(knitr)`; otherwise, you may get errors like "object 'opts_chunk' not found";

- The chunk options `fig.path` (figure path) and `cache.path` (cache path) are modified in **rmarkdown** when rendering an Rmd file. In **knitr**, they are `figure/` and `cache/`, respectively. Now in **rmarkdown**, they are `foo_files/figure-format/` and `foo_files/cache-format/`, respectively, where `foo` is the base filename of the input Rmd file without the file extension, and `format` is the output format, e.g., `tex` or `html`;

- The chunk option `error` was changed from `TRUE` to `FALSE`, and the implication is that R will stop by default, instead of showing the error messages in the R Markdown output document (see Section 6.2.4);

- The chunk options `fig.width`, `fig.height`, and `fig.retina` may take different values, depending on the output format. You can either check the **rmarkdown** documentation of output format functions, or print `str(knitr::opts_chunk$get())` in your R Markdown document to see the values of chunk options.

# 15

## Applications

So far we have been introducing the usage of **knitr** with short examples for the sake of simplicity. In this chapter we use some concrete and complete examples to show how **knitr** works with real applications; we do not explain every single detail of these applications, and we only point out the critical parts in them.

## 15.1 Homework

For homework applications, R Markdown might be the preferred document format to work with due to its simplicity, and homework is usually not targeted at publication. As mentioned before, RPubs (http://rpubs.com) is a platform for sharing (HTML) reports generated from RStudio by **knitr**. There are many homework submissions, too.

Since a homework report is relatively simple, we may not need too many **knitr** features; some common features used in homework are: set the size of plots (fig.width and fig.height), hide the source code because the grader may not wish to read it (echo = FALSE), and enable cache for time-consuming computing jobs (cache = TRUE), etc. Other features that come by default such as tidy = TRUE and highlight = TRUE can help users who do not care about coding styles produce more readable code in the output document.

Now we show an example of Gibbs sampling. For the bivariate Normal distribution

$$\begin{bmatrix} X \\ Y \end{bmatrix} \sim \mathcal{N} \left( \begin{bmatrix} \mu_X \\ \mu_Y \end{bmatrix}, \begin{bmatrix} \sigma_X^2 & \rho\sigma_X\sigma_Y \\ \rho\sigma_X\sigma_Y & \sigma_Y^2 \end{bmatrix} \right) \qquad (15.1)$$

we know the conditional distributions

$$Y|X = x \sim \mathcal{N} \left( \mu_Y + \frac{\sigma_Y}{\sigma_X}\rho(x - \mu_X), (1 - \rho^2)\sigma_Y^2 \right)$$

$$X|Y = y \sim \mathcal{N} \left( \mu_X + \frac{\sigma_X}{\sigma_Y}\rho(y - \mu_Y), (1 - \rho^2)\sigma_X^2 \right) \qquad (15.2)$$

so we can use the Gibbs sampling to generate random numbers from the joint Normal distribution. First we initialize $x^{(0)}$ and $y^{(0)}$, then repeatedly generate $x^{(k)} \sim f(x|y^{(k-1)})$ and $y^{(k)} \sim f(y|x^{(k)})$. The R code below is a translation of 15.2:

```r
rbinormal <- function(n, mu1, mu2, sigma1, sigma2, rho) {
    # initialize
    x <- rnorm(1, mu1, sigma1)
    y <- rnorm(1, mu2, sigma2)
    xy <- matrix(nrow = n, ncol = 2, dimnames = list(NULL,
        c("X", "Y")))
    # sample from conditional distributions
    for (i in 1:n) {
        x <- rnorm(1, mu1 + sigma1/sigma2 * rho * (y - mu2),
            sqrt(1 - rho^2) * sigma1)
        y <- rnorm(1, mu2 + sigma2/sigma1 * rho * (x - mu1),
            sqrt(1 - rho^2) * sigma2)
        xy[i, ] <- c(x, y)
    }
    xy
}
```

Figure 15.1 shows the first 20 steps of Gibbs sampling for the bivariate Normal distribution with $\mu_X = 0$, $\sigma_X = 2$, $\mu_Y = 1$, $\sigma_Y = 3$, $\rho = 0.7$.

```r
set.seed(123)
n <- 20
z <- rbinormal(n, mu1 = 0, mu2 = 1, sigma1 = 2, sigma2 = 3,
    rho = 0.7)
plot(z, pch = 19)
arrows(z[-n, 1], z[-n, 2], z[-1, 1], z[-1, 2], length = 0.15,
    col = "gray40")
```

And we can draw some samples as well:

```r
z <- rbinormal(5000, 0, 1, 2, 3, 0.7)
smoothScatter(z, nbin = 64)
points(0, 1, col = "white", pch = 19)   # theoretical mean
```

Figure 15.2 shows 5,000 samples from this distribution, and we can calculate the sample means, standard deviations, and the correlation, which should be close to the corresponding theoretical values:

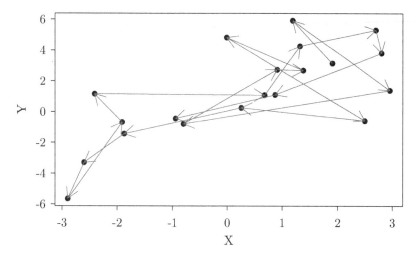

FIGURE 15.1: Trace of Gibbs sampling for a bivariate Normal distribution: the arrows show the first 20 steps of Gibbs sampling.

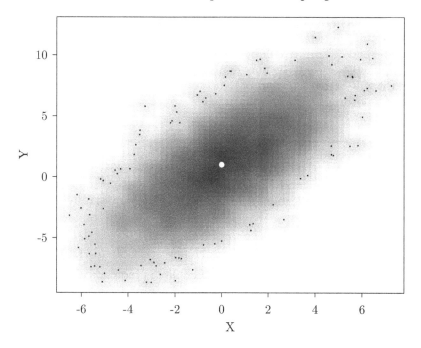

FIGURE 15.2: 5000 points from Gibbs sampling: the smoothed scatterplot shows the density of the 2D distribution.

```
apply(z, 2, mean)  # sample mean
```

```
##        X        Y
## 0.001287 0.971010
```

```
apply(z, 2, sd)  # sample sd
```

```
##     X     Y
## 1.973 2.971
```

```
cor(z)  # sample correlation
```

```
##        X        Y
## X 1.0000 0.6948
## Y 0.6948 1.0000
```

In this small application, we used cache (although this particular
example is not too slow) and TikZ graphics. We adjusted the plot sizes
(5 × 3 for Figure 15.1 and 5 × 4 for Figure 15.2). Note the narratives and
code chunks are interwoven, and the reader can learn the theory, see
the computing, and verify the results in the same report. Everything
is transparent, and it will be easy to find out errors. Sometimes the
computer code we write may not really reflect what we said in theory,
and it will be hard to find out such errors if we separate computing
from reporting.

In terms of data, code and software sharing, we cannot
yet rely on goodwill and self discipline when it comes to
sharing publication material and making studies fully reproducible.

Huang and Gottardo (2013)

Comparability and reproducibility of biomedical data

People have been proposing sharing data, code, and software in
data analysis for the sake of reproducible research, e.g., Huang and
Gottardo (2013). We believe that more efforts in education should be
an important step, and we can start with reproducible homework.

## 15.2   Serve Dynamic Documents

The **servr** package (Xie, 2015c) provides some simple HTTP server functions to serve files under a given directory based on the **httpuv** package. To some degree, this package is like python -m SimpleHTTPServer or python -m http.server if you are familiar with Python. Originally it was designed to serve static files under a directory, and the main function was *httd()*:

```
servr::httd("./")
```

If you run the above function in the R console, R will launch your Web browser to show a list of files under the current working directory (./), or show index.html if this file exists. You can click the links on the files to view their content.

Later **servr** was extended based on **knitr** and **rmarkdown**, so it can also serve dynamic R Markdown documents. There are functions *jekyll()*, *rmdv1()*, and *rmdv2()* in this package to serve HTML files generated from R Markdown documents (via **knitr** or **rmarkdown**). R Markdown documents can be automatically recompiled when their HTML output files are older than the corresponding source files, and HTML pages in the Web browser can be automatically refreshed accordingly, so you can focus on writing R Markdown documents, and results will be updated on the fly in the Web browser. This saves you two steps: click the Knit HTML button, and refresh the Web browser. Both steps can be distracting when you write a report. With **servr**, all you need to do is write the R Markdown document after you launch a server.

This is even more useful when you write R Markdown documents in the RStudio IDE, because **servr** has set the Web browser to be the RStudio Viewer by default when it detects the RStudio IDE, and you can put the source document and its output side by side like the layout in Figure 15.3. It is completely fine if you do not use RStudio — the automatic compilation and refreshing also work if you use other editors and Web browsers.

The functions *rmdv1()* and *rmdv2()* correspond to R Markdown v1 and v2, respectively. After you call servr::rmdv1() or servr::rmdv2() in the R console, you can click the HTML file foo.html if it has its source document foo.Rmd, and view the HTML output. Then whenever you edit foo.Rmd and save it, **servr** will automatically recompile it and refresh the HTML output page.

The function *jekyll()* is like *rmdv1()* and *rmdv2()*, but is tailored for Jekyll websites. We have briefly introduced Jekyll in Section 13.4. It

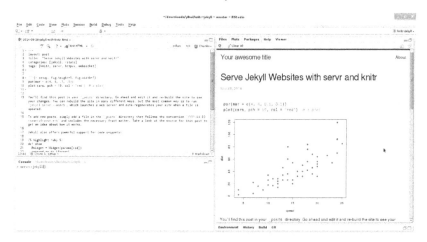

FIGURE 15.3: The layout of an R Markdown document (top-left panel) and its output in the RStudio Viewer (right panel): we typed a **servr** function in the R console (bottom-left), and the output of the R Markdown is showed in the RStudio Viewer. This figure is only for illustration purposes; see `https://github.com/yihui/servr` for the original image if you want to read the text in it.

is tedious to compile R Markdown posts or pages to Markdown again and again, and that is why *jekyll()* can be useful. Once you call the function `servr::jekyll()` in the root directory of a Jekyll website, you will get a preview of the website in your Web browser. Besides, as you edit and save your blog post, the Web browser will refresh the page to show the updated output. The **knitr-jekyll** repository (`https://github.com/yihui/knitr-jekyll`) is an example of serving Jekyll websites using **servr**.

Later we will introduce package vignettes in Section 15.4, and the function *vign()* in **servr** can be used to serve HTML vignettes while we develop an R package. Its advantage is that it does not preserve the HTML output file in the source package when serving the vignette, which makes the source package clean.

For those who are curious about the technical details, the implementation is based on WebSockets. When **servr** shows an HTML page, it also injects a piece of JavaScript code in it to set up a WebSocket connection to talk to R periodically (e.g., on one-second basis). Every time R receives a request from the WebSocket, it will compare the timestamps of Rmd files with their output HTML files. If an Rmd file is newer than its HTML output, **servr** will call **knitr** or **rmarkdown** to recompile the Rmd file to HTML, then send a message back to the WebSocket.

```
all: example.html

%.html: %.Rmd
    Rscript -e "rmarkdown::render('$^')"
```

FIGURE 15.4: A Makefile example for the function *make()* in **servr**: the HTML file to be generated is specified in the target all, and a rule is specified on how to generate an HTML file from an Rmd file via **rmarkdown**.

When the WebSocket receives this message, it calls location.reload() in JavaScript to refresh the page.

A critical step in this process is to check if we need to recompile any Rmd files. This is a task that GNU Make (http://www.gnu.org/software/make/) is good at, so **servr** also provided a function *make()* so that you can provide your own Makefile to rebuild Rmd files when necessary. Figure 15.4 is an example Makefile for the *make()* function.

By default, a server function will block the current R session, which can be a problem if you want to continue working in the same R session. To solve this problem, you can use the argument daemon = TRUE for the server function, e.g., httd(daemon = TRUE), or rmdv2(daemon = TRUE). This tells **servr** to launch a daemonized server that will not block the current R session.

## 15.3   Website and Blogging

We introduce a few websites and blogs built upon **knitr** in this section, and the Web pages are created from either R Markdown or R HTML.

### 15.3.1   Vistat and Rcpp Gallery

Vistat (http://vis.supstat.com) is a website based on R Markdown and Jekyll (Section 13.4). It aims to provide a gallery of reproducible statistical graphics. The repository for the website is publicly available on Github: https://github.com/supstat/vistat.

The core of this repository is the R script ./_bin/knit, which sets some global chunk options and compiles Rmd documents to Markdown output. Math equations are rendered by MathJax, animations are supported through the **SciAnimator** library (Section 7.3.1), and we can also create Web graphics via the **D3** library.

After **knitr** has compiled Rmd source files to Markdown files, Jekyll can compile Markdown to HTML, which gives us a website.

The Rcpp Gallery (`http://gallery.rcpp.org`) is a website for **Rcpp** (Eddelbuettel et al., 2015) articles and examples, and it is also built on R Markdown; in particular, it uses **knitr**'s Rcpp engine (Section 11.2.1).

### 15.3.2   UCLA R Tutorial

The UCLA Statistical Consulting Group has maintained software tutorials for several statistical packages for many years, and one of them is dedicated to R: `http://www.ats.ucla.edu/stat/r/`. Before 2012, this website was built by cut-and-paste. The results were generated in R and copied into the HTML pages. After **knitr** was released in 2012, one of the Web administrators, Joshua Wiley, decided to rewrite the R tutorial pages with **knitr** instead of using the R HTML format. Now it is much easier to maintain the Web pages, and the R output also has much better reproducibility. After R is updated or any dataset is changed, the whole website can be rebuilt automatically by compiling all source documents again.

### 15.3.3   The cda and RHadoop Wiki

Github has an integrated Wiki system for each repository. We can write wiki pages in a variety of formats, such as Markdown and reStructured-Text, etc. Each page is essentially a file, and the wiki is essentially a Git repository; therefore we can write Rmd files and compile them to Markdown files, and push to Github through Git.

The **cda** package (Auguie, 2013) used the above approach to build its wiki site on Github: `https://github.com/baptiste/cda/wiki`. We can find the Rmd source files under the wiki directory of the package.

The RHadoop project has a similar wiki at `https://github.com/RevolutionAnalytics/RHadoop/wiki`.

### 15.3.4   The ggbio Package

The **ggbio** package (Yin et al., 2012) is an R implementation for extending the Grammar of Graphics for genomic data based on the **ggplot2** package. It has a website, `http://tengfei.github.com/ggbio/`, on which we can find its documentation. The function *knit_rd()* (Section 12.4.8) was used to compile its R documentation pages to HTML, so we can directly see the output of the examples. Once this package has been installed, it only needs one line of code to get the HTML pages:

```
knitr::knit_rd("ggbio")
```

Then we can publish the HTML files to Github, and we do not need to do anything with the images because they are base64 encoded in the files.

By the way, the **ggbio** package also has a PDF vignette written with **knitr**, which can be found on the website or with the command:

```
vignette("ggbio", package = "ggbio")
```

### 15.3.5   Geospatial Data in R and Beyond

Barry Rowlingson gave a tutorial workshop on geospatial data analysis in R at the useR! 2012 conference, and here is the corresponding website: http://www.maths.lancs.ac.uk/~rowlings/Teaching/ UseR2012/. The website was created from R HTML files and has a nice style from Twitter Bootstrap (a popular CSS framework). The advantage of using R HTML over R Markdown is that we have full control of the style; this website is a good example of arranging R code chunks and output in `div` elements with custom CSS styles.

## 15.4   Package Vignettes

As discussed by Gentleman and Temple Lang (2004), R packages have the great potential of building and disseminating reproducible reports, besides their obvious functionality of providing computing routines. Specifically, R package vignettes can be an ideal format for writing reproducible reports, with other components of the package providing the infrastructure such as functions, unit tests, and datasets. An R package vignette is just like a paper, and the output is dynamically compiled from its source document during the package building process, i.e., `R CMD build`.

For R under the version 3.0.0, it uses Sweave to build package vignettes. Due to the limitations of Sweave (Section 16.1) and the barrier of LATEX, R package vignettes were not widely used before R 3.0.0. Bio-Conductor is an exception, though, because vignettes are mandatory for packages on BioConductor.

It has become much more natural and easy to compile package vignettes since R 3.0.0, thanks to Henrik Bengtsson, Duncan Murdoch, and R core. Now there are more than 500 package vignettes compiled from **knitr** in about 300 packages on CRAN (`https://gist.github.com/yihui/7698648`). In the next section, we introduce **knitr** vignette engines, and then we show a few examples. Sections 15.4.3 and 15.4.4 are only for those who are interested in older versions of R, and we no longer recommend that you use the tricks mentioned in these two sections.

### 15.4.1   Vignette Metadata and Engines

To use **knitr** to build vignettes, we only need to follow these simple steps:

- specify a vignette engine, such as `%\VignetteEngine{knitr::knitr}`, in the vignette source document (e.g., an Rnw or Rmd file)

- add a field `VignetteBuilder: knitr` in the package DESCRIPTION file

- add `knitr` to the `Suggests` field in DESCRIPTION

Then we can write vignettes using the **knitr** syntax (e.g., `<<>>=` or ```` ```{r} ```` for code chunks). Remember vignettes are put under the vignettes/ directory of the package root directory.

According to the R manual "Writing R Extensions," we also have to write the title of the vignette in `\VignetteIndexEntry{}`. There are a few other optional metadata specifications such as `\VignetteKeyword{}`. See Figure 15.5 for an example of the vignette metadata (title and vignette engine) for an R Markdown v2 vignette in **knitr**. After we build the package, the vignettes will be listed in an HTML index page.

The **knitr** package has several PDF and HTML vignettes compiled in this way, and we can view them by running:

```
browseVignettes(package = "knitr")
# or view specific vignettes if you know their filenames
vignette("knitr-intro", package = "knitr")
vignette("knitr-refcard", package = "knitr")
```

The vignette engine `knitr::knitr` is only one of the possible engines in **knitr**. To see all of them, you can use the function *vignetteEngine()* in the **tools** package:

```
---
title: "Not An Introduction to knitr"
author: "Yihui Xie"
date: "`r Sys.Date()`"
bibliography:
  - ../inst/examples/knitr-packages.bib
  - ../inst/examples/knitr-manual.bib
vignette: >
  %\VignetteEngine{knitr::rmarkdown}
  %\VignetteIndexEntry{Not an Introduction to knitr}
output: knitr:::html_vignette
---
```

FIGURE 15.5: The metadata of a **knitr** vignette: this is extracted from the **knitr** vignette, and you can find it from `system.file('doc', 'knitr-intro.Rmd', package='knitr')`.

```
library(knitr)
sort(names(tools::vignetteEngine(package = "knitr")))

## [1] "knitr::docco_classic"
## [2] "knitr::docco_classic_notangle"
## [3] "knitr::docco_linear"
## [4] "knitr::docco_linear_notangle"
## [5] "knitr::knitr"
## [6] "knitr::knitr_notangle"
## [7] "knitr::rmarkdown"
## [8] "knitr::rmarkdown_notangle"
```

The engines with the suffix _notangle have the same weave functions as those without the suffix, but have disabled the tangle function, meaning that there will not be R scripts generated from vignettes during R CMD build or R CMD check. Sometimes we may not want to tangle R scripts from vignettes, because it is redundant for R CMD check to run the same code again after the code has been executed in weave, and currently the inline R code expressions are not included in the tangle output, which can also cause problems.

Please note the :: operator has no special meaning in a vignette engine. It can be misleading because :: is an operator in base R that fetches an exported object from a package, e.g., stats::lm. However, in the vignette engine notation, :: is nothing but a delimiter that separates the package name from the engine name, so knitr::rmarkdown does not mean `rmarkdown` is a function in **knitr**, but only one of the vignette engines in **knitr**.

When you use the `rmarkdown` vignette engine, you are free to choose the output format, as long as the filename extension is .html or .pdf, because R only recognizes these two types of vignette output at the moment. When the output format is HTML, it can be an HTML document, or any of the HTML5 presentations (e.g., ioslides or Slidy). When it is PDF, it can be either a PDF document or Beamer slides.

### 15.4.2   Vignette Examples

We have put together a list of vignettes from current CRAN packages using the **knitr** vignette engines at `https://gist.github.com/yihui/7698648`, and you can learn from these examples.

The **ggplot2** transition guide by Murphy (2012) is a great example of an R package vignette, although it is not shipped with the **ggplot2** package. This guide was intended to announce new features and explain changes in **ggplot2** 0.9.0, which may affect users of older versions.

One nice feature of this guide is that we can compile the Rnw document to either a color or a black/white version, which is controlled by a global variable `bw_version`; if it is `TRUE`, a black and white version will be produced. This is achieved by setting the chunk options `eval = bw_version` and `echo = bw_version` for the chunks that produce black/white plots, and in **ggplot2** this means *theme_bw()* and gray scales such as *scale_fill_gray()*. When `bw_version` is `FALSE`, these chunks will be hidden from the output (the source code is neither evaluated nor echoed). Similarly, there are some other chunks that have the options `eval = !bw_version` and `echo = !bw_version`, and these chunks produce color plots. In all, we can control if the PDF output is color or black/white by a single variable, which is very convenient (recall Section 5.1.1). Figure 15.6 is a sample page of the transition guide from the color version.

The **corrplot** package (Wei, 2013) has an example of HTML vignettes. You can find the source document of its vignette on Github at `https://github.com/taiyun/corrplot/tree/master/vignettes`. Obviously, it is an Rmd document (Section 5.2.1). Note it uses R Markdown v1. Open it with a text editor (e.g., RStudio) and we will see R code chunks in it. We can view the HTML vignette compiled from it in the Web browser by running:

```
help(package = "corrplot", help_type = "html")
```

This shows the HTML index page of the **corrplot** documentation,

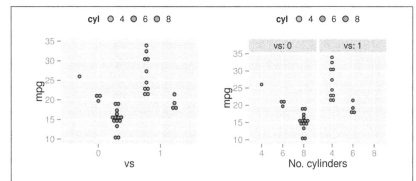

## 3.4 geom_violin()

This function generates violin plots in **ggplot2**, a way to plot one or more continuous density estimates that is particularly useful when comparing multiple groups. A violin plot is a combination of a box plot and a kernel density estimate, the latter of which is rotated to run alongside the box plot symmetrically on each side. The examples below come from the function's help page.

In geom_violin(), violins are automatically dodged when any aesthetic is a factor. By default, the maximum width is scaled to be proportional to the sample size. In the plot on the far right below, the bandwidth of the kernel density estimator is reduced from the default 1, which makes for a less smooth density estimate and hence a less smooth violin plot.

```
p <- ggplot(mtcars, aes(factor(cyl), mpg))
p + geom_violin()    # default scale is "count"
p + geom_violin(aes(fill = factor(cyl), colour = factor(cyl)))
p + geom_violin(adjust = 0.5)
```

The next set of plots simply play around with a few extra features. The plot on the left adds a strip plot to the violin for each group. The central plot adds fill color and alpha transparency to the violins and is augmented with boxplots. The plot on the far right adds a dot plot around

FIGURE 15.6: A sample page of the **ggplot2** transition guide: introducing the new geom added to **ggplot2** 0.9.0 — *geom_violin()*.

```
PDFS= foo.pdf bar.pdf

all: $(PDFS)

clean:
    rm -f *.tex *.bbl *.blg *.aux *.out *.log

%.pdf: %.Rnw
    $(R_HOME)/bin/Rscript -e "knitr::knit2pdf('$*.Rnw')"
```

FIGURE 15.7: The Makefile to compile PDF vignettes using **knitr**: use *knit2pdf()* to compile Rnw documents to PDF.

and we can see the link to the vignette "Overview of user guides and package vignettes." Since **corrplot** is a package for visualizing correlation matrices, it has many graphical examples, which are shown in its HTML vignette.

The source package of **knitr** contains a mixture of PDF and HTML vignettes, all of which are listed in the HTML help page of this package.

The **sampSurf** package (Gove, 2013) also has a nice HTML vignette at http://sampsurf.r-forge.r-project.org, which was created from an R HTML source document and even contains some 3D plots produced by the **rgl** package.

### 15.4.3   PDF Vignette

If we want to build vignettes with **knitr** for R <= 3.0.0, we have to use some tricks. One way to do this is through a Makefile (http://www.gnu.org/software/make/), which will be used by R CMD build when building vignettes. In this Makefile, we can set our rules to create the PDF file using a custom tool like **knitr**.

The Makefile is under the vignettes/ directory in the source package. When R compiles vignettes, it calls *Sweave()* first; if there is a Makefile, the make command will be run on it. In the Makefile, we also have access to R, so it is possible to call **knitr** via command line to compile vignettes. Figure 15.7 shows a sample of the Makefile to be used to compile vignettes with **knitr**. The key is to run knitr::knit2pdf() on the Rnw files; we put all PDF files to be generated in the variable PDFS.

Obviously, the disadvantage of this approach is that all Rnw documents have to be compiled by Sweave before any further processing.

```
HTMLS= foo.html bar.html

all: $(HTMLS)

clean:
    rm -rf figure/ *.md

%.html: %.Rmd
    $(R_HOME)/bin/Rscript -e "knitr::knit2html('$*.Rmd')"
```

FIGURE 15.8: The Makefile to compile HTML vignettes: use *knit2html()* to compile Rmd documents to HTML.

Besides, the new approach in R >= 3.0.0 does not require the make utility to be installed.

### 15.4.4   HTML Vignette

Similarly, we can create package vignettes in the HTML format from R Markdown documents. Again, the HTML vignettes had to be compiled by a Makefile before R 3.0.0. Figure 15.8 shows the source of a sample Makefile for building HTML vignettes, where the function *knit2html()* was called. Note make clean will remove the figure/ directory, which is due to the fact that images generated by **knitr** will be base64 encoded in the HTML output, so the image files are no longer needed.

## 15.5   Books

We can also write books with **knitr**. At the time of writing this book, at least one book has been published (Lebanon, 2012), and the book *Regression Modeling Strategies* (Harrell, 2001) is under revision for a new edition, which is based on **knitr**.

### 15.5.1   This Book

In the spirit of "eating one's own dog food" (see Wikipedia if this is unclear), this book was written with **knitr** in LyX (see Section 4.2). The whole book is in one L<sub>Y</sub>X file, although it is entirely possible to split chapters into individual files.

A few chunk options were set globally in the very beginning of the document, such as `cache = TRUE` (for speed), `dev = 'tikz'` (for style of graphics), and `fig.align = 'center'` (for alignment of plots). We also set `options(formatR.arrow = TRUE)` (see the **formatR** package), because the author's preference of the assignment operator is = instead of <-, but <- is more commonly used by R users; this option allows the equal signs to be replaced by the left arrows automatically wherever applicable, although all I typed are actually equal signs.

We have a few chunk hooks (Chapter 10) in this book for various purposes. For example, there is a par hook that sets the graphical parameters to this:

```
par(mar = c(4, 4, 0.1, 0.1), cex.lab = 0.95, cex.axis = 0.9,
    mgp = c(2, 0.7, 0), tcl = -0.3, las = 1)
```

So when we want to use this set of parameters, we just add a chunk option `par = TRUE` instead of having to type it again and again.

Although we see the code chunks and the plots are separate in this book, that is not true in the source document: the code chunks are actually inside the figure environments, but we used the document hook *hook_movecode()* to move code chunks out of the figure environments eventually.

Because we have to show chunk headers occasionally for pedagogical purposes, we have a chunk hook named append to add <<>>= and @ to the chunk output:

```
knit_hooks$get("append")

## function(before, options, envir) {
##     txt = options$append[[ifelse(before, 1, 2)]]
##     txt = c("\\begin{alltt}", txt, "\\end{alltt}")
##     paste(txt, collapse = "")
## }
```

Basically this hook enables us to write additional character strings before and/or after a chunk; e.g., we can use the chunk option append = list('<<A>>=', '@') to add the syntax information to the chunk output. We need to use this hook because we cannot write the chunk headers directly in the source document, otherwise they will be parsed and disappear in the final output.

There is an output hook that modifies the default plot hook function by adding a frame box to a plot, and it was used in Figure 10.3 and Figure 10.4.

The bibliography database of all R packages is dynamically written by the *write_bib()* function as introduced in Section 12.4.1, so it is guaranteed that the version information is up to date (at least before the manuscript was submitted to the publisher).

### 15.5.2 The Analysis of Data

Another notable example is the book *The Analysis of Data* by Lebanon (2012); the most notable feature of this book is that it has the double PDF/HTML versions. The HTML version is freely available at http://theanalysisofdata.com. Both versions are produced from essentially the same set of source documents. For the HTML version, there are additional settings, for example, the typesetting of math equations is done by the MathJax library, so it has to be included in the head section of the HTML source.

### 15.5.3 The Statistical Sleuth in R

*The Statistical Sleuth* (Ramsey and Schafer, 2002) is an excellent text in statistics, and one feature of this book is that it has a large number of datasets. The book itself was not written with **knitr**, but some other authors (Horton et al., 2012) have created a website (http://www.math.smith.edu/~nhorton/sleuth/) in which they re-did a lot of the data analysis examples in the book in R. You can check out both the PDF documents and the Rnw source files on the website.

### 15.5.4 Text Analysis with R for Students of Literature

The book *Text Analysis with R for Students of Literature* by Jockers (2014) was written using LATEX and **knitr**. The most amazing fact about this book is perhaps that its author taught himself LATEX before he started putting together this book in LATEX, and finished the book draft in just a couple of months. The book is an introduction to computational text analysis, and has a lot of short examples. It would be extremely tedious if the author had to run each example and copy the output to the LATEX manuscript by hand.

## 15.6   Literate Programming for R Packages

Although we have introduced Literate Programming (LP) in the beginning of this book, we do not actually use the **knitr** package for *programming* purposes. Most of the time we use **knitr** for data analysis and reporting purposes instead. The original LP paradigm is about both weaving and tangling: we may weave a source document to software documentation, or tangle the program code to execute it. Apparently, we do not really have to tangle the program code for execution purposes when using **knitr**, because code execution occurs right in the process of weaving.

Interestingly, the most common application of Knuth's original LP paradigm seems to be documenting software (using a special form of comments) for users instead of "programming" for package authors. In other words, we use LP to document the *usage* of software, instead of documenting the *source code*. See Doxygen (van Heesch, 2008), Javadoc (http://en.wikipedia.org/wiki/Javadoc), and **roxygen2** (Wickham et al., 2015) for examples. There exists one exception, though, in the LATEX world. Some LATEX package authors write both LATEX code and documentation in a single document, and weave it into a PDF document that contains both the source code and documentation. This is not entirely surprising, considering Knuth's original implementation of LP using TEX and Pascal. There is a small number of R packages using LP as well, such as Terry Therneau's **survival** and **coxme** packages.

LP does not seem to be a popular approach to programming, but it is still an interesting idea, and can be useful especially when it is applied to your own favorite language. It may be boring for some people to read LATEX source code, but reading R source code can be more pleasant. Objective opinions aside, we believe LP has at least two advantages:

1. You can write much more extensive and richer documentation than you normally could do with comments. In general, comments in code are (or should be) brief and limited to plain text. Normally you will not write five paragraphs of comments to explain a few lines of code, and you cannot write readable math expressions or embed a video in comments.

2. You can label code chunks and reference/reuse them using the labels, which allows you to compose your program flexibly using different pieces of code chunks. For example, you can define and explain a code chunk later in the document, but insert it in a previous code chunk using its label. This feature has been emphasized by Knuth, but it is not widely

adopted for some reason. Perhaps most people are more comfortable with designing a big program by smaller units like functions instead of code chunks, which is actually a good idea.

In fact, we can apply LP to developing R packages. There are multiple ways to achieve the goal, and we only introduce one here, using the following tools:

1. The *purl()* function in **knitr**, which makes it possible to extract program code from a source document;
2. Package vignettes, which can contain both program code and documentation;
3. GNU Make, which allows us to define when and how to generate an output file from a source file.

The **rlp** package (`https://github.com/yihui/rlp`) is an example of writing an R package using LP techniques. You can find details in this repository, and the basic idea of the implementation is:

1. Instead of writing R source code under the R/ directory of the package, we can write the code in package vignettes (R Markdown) under the vignettes/ directory;
2. Use a Makefile to define how to generate R scripts R/*.R from vignettes vignettes/*.Rmd;
3. Run `make` to generate R scripts to R/ and `R CMD build` to build the package.

These steps can be made easy by using the RStudio IDE, and we can actually just click a button to do the these steps. The implementation details are too technical and specific for this book, and we will leave it to the readers to go through the documentation of this package.

# 16

## Other Tools

Besides **knitr**, there is a large number of other tools for dynamic documents. Some are R packages, and others are tools in other languages such as Python and awk. We give a brief overview of these tools with comparisons to **knitr** in this chapter, and we especially explain the differences between Sweave and **knitr** for Sweave users.

### 16.1  Sweave

The **knitr** package was largely motivated by Sweave (Leisch, 2002), which has been a longstanding prominent tool for dynamic documents in R, and is a part of base R (in the **utils** package as the *Sweave()* function). Sweave primarily deals with Rnw documents, although it also has a modular design that allows it to be extended to other document formats. A number of extensions based on Sweave exist on CRAN, and we will introduce them in the next section.

There are two ways to run Sweave. We can call it in an interactive R session (you do not need to load the **utils** package):

```
Sweave("your_file.Rnw")  # gives you your_file.tex
```

In addition, we can use the command line, too:

```
R CMD Sweave your_file.Rnw
```

Since Sweave is part of base R, its development has almost plateaued in recent years. Another major problem is that its modular design is not modular enough, so its extensions may become incompatible as Sweave gets updated in base R. As far as we know, a few R packages based on Sweave copied a large amount of core code from Sweave, and are no longer synchronized with the development of Sweave.

A lot of **knitr**'s chunk options were borrowed from Sweave, such

233

as `eval`, `echo`, `results` and so on, but the design is different, so there are several differences between them. Before version 1.0, **knitr** tried to be compatible with Sweave — **knitr** was able to compile Sweave documents because of some internal functions to fix the differences automatically. The compatibility has been dropped since v1.0, with a conversion function *Sweave2knitr()* provided to convert Sweave documents to **knitr** manually. Below is an example of converting the Rnw document in the **utils** package and showing the differences after conversion (< shows the original document, and > shows the converted file):

```
testfile <- system.file("Sweave", "Sweave-test-1.Rnw",
    package = "utils")
outfile <- tempfile(fileext = ".Rnw")
Sweave2knitr(testfile, output = outfile)

# capitalizing true/false to TRUE/FALSE:
#     * fig=true
# removing the unnecessary option fig=TRUE:
#     * fig=TRUE
#     * fig=TRUE
# quoting the results option:
#     * results=hide
# removing options 'print', 'term', 'prefix':
#     * print=TRUE
#     * echo=TRUE,print=TRUE
# capitalizing true/false to TRUE/FALSE:
#     * echo=true
# changing \SweaveOpts{} to opts_chunk$set():
#     * \SweaveOpts{echo=FALSE}
#     * \SweaveOpts{echo=true}
# removing extra lines (#n shows line numbers):
#     * (#69) @

cat(system(sprintf("diff %s %s", shQuote(testfile),
    shQuote(outfile)), intern = TRUE), sep = "\n")

# 7c7,14
# < \SweaveOpts{echo=FALSE}
# ---
# >
# > <<include=FALSE>>=
# > library(knitr)
# > opts_chunk$set(
# > echo=FALSE
```

```
# > )
# > @
# >
# 15c22
# < <<print=TRUE>>=
# ---
# > <<>>=
# 17c24
# < <<results=hide>>=
# ---
# > <<results='hide'>>=
# 22c29
# < <<echo=TRUE,print=TRUE>>=
# ---
# > <<echo=TRUE>>=
# 43c50,57
# < \SweaveOpts{echo=true}
# ---
# >
# > <<include=FALSE>>=
# > library(knitr)
# > opts_chunk$set(
# > echo=TRUE
# > )
# > @
# >
# 53c67
# < <<fig=TRUE>>=
# ---
# > <<>>=
# 63c77
# < <<fig=true>>=
# ---
# > <<>>=
# 69d82
# < @
```

### 16.1.1  Syntax

By default, **knitr** uses a new type of syntax to parse chunk options, which is similar to R function arguments. This gives us much more power than the traditional Sweave syntax. We can use arbitrary objects in chunk options and make use of the full power of R.

Sweave treats chunk options as character strings and parses them by splitting the options by commas, whereas **knitr** uses the R syntax: if the option takes a character value, we have to quote it just like we do in R, e.g., results = 'hide' (in Sweave we write results = hide). See Section 12.1.3 for an example of doing computing directly in chunk options. Below is another example, which shows how flexible the new syntax is (we can dynamically create a figure caption):

```
<<cap, fig.cap=paste('The P-value is', t.test(x)$p.value)>>=
x <- rnorm(100)
boxplot(x)
@
```

The other minor difference in syntax is that **knitr** does not recognize @ as the beginning of text chunks unless there is a chunk header before it. For example, **knitr** will keep the first @ in the example below but Sweave will remove it:

```
text
@
<<A>>=
1 + 1
@
```

*Sweave2knitr()* can fix this problem automatically.

### 16.1.2 Options

Some options of Sweave were dropped in **knitr** and some were changed, including:

**concordance** was changed mainly to support RStudio; if the package option opts_knit$get('concordance') is TRUE, a file named input-concordance.tex will be written with output line numbers mapped to input line numbers; note the implementation is less accurate than Sweave

**keep.source** was merged into a more flexible option tidy

**print** was dropped: whether an R expression is going to be printed is consistent with your experience of using R (e.g., x <- 1 will not be printed, while 1:10 will; just imagine you are typing the commands in an R console); if you really want the output of an expression to be invisible, you may use the function *invisible()*

**term** was dropped (think `term = TRUE`)

**prefix** was dropped (think `prefix = TRUE`)

**prefix.string** was renamed `fig.path` and it is always used for figure filenames

**eps, pdf** and all logical options for graphics devices were dropped: use the new option `dev` instead, which is similar to `grdevice` in Sweave but has more than 20 predefined graphical devices; see Chapter 7

**fig** was dropped; now use `fig.keep`: `fig.keep = 'high'` in **knitr** is equivalent to `fig = TRUE` and `fig.keep = 'none'` is the same as `fig = FALSE` in Sweave

**width, height** were renamed `fig.width` and `fig.height`, respectively

Meanwhile, `\SweaveOpts{}` and `\SweaveInput{}` are deprecated; use `opts_chunk$set()` and the chunk option `child` to set global chunk options and include child documents, respectively.

For logical options, only `TRUE/FALSE/T/F` are supported (the first two are recommended), and `true/false` will not work; e.g., `eval = FALSE` is OK, and `eval = false` is not (unless there is an R object named `false` that happens to take a logical value `FALSE`). Chunk reference using the `<<label>>` syntax is still available, and there are other approaches for reusing chunks, e.g., use the new option `ref.label`; chunk references can be recursive, as introduced in Chapter 9.

### 16.1.3 Problems

Some known problems and frequently asked questions in Sweave have been solved in **knitr**:

- empty figure chunks give LATEX errors in Sweave but not in **knitr** because figures will not be generated at all; **knitr** writes figures to LATEX only when there are plots in a chunk

- **lattice** (and **ggplot2**) graphics do not work in Sweave if you do not explicitly *print()* them, and they work in **knitr** just like in R console (if these plot objects appear in the top environment, you do not need to print them)

- the width of figures in the output is set to `.8\textwidth` in Sweave by default via `\setkeys{Gin}{width=.8\textwidth}` defined in the LATEX style Sweave.sty; this affects all figures in the document regardless of whether they are generated by Sweave, and there is no straightforward way to set individual widths for figures; this problem has been solved by the `out.width` option in **knitr**

- multiple figures from one figure chunk do not work by default in Sweave and you have to write LATEX code by yourself in this case; for **knitr**, it does not make any difference no matter how many plots there are in one chunk

- it is possible to use output hooks to change the formatting of output in **knitr**, and we do not have to use hard-coded LATEX environments such as Sinput/Soutput in Sweave; in fact, we can call *render_sweave()* to render the Sweave style from **knitr**

- it is easy to produce HTML output with **knitr** (with either R HTML or R Markdown), and Sweave needs extensions such as **R2HTML**, which only deals with HTML

Sometimes we see a stray Rplots.pdf file after we run Sweave, and that is because R's default graphical device is *pdf()* for non-interactive R sessions, which creates Rplots.pdf. In **knitr**, the default device is set to a null device (pdf(file = NULL)) so that no stray PDF files will be generated.

## 16.2   Other R Packages

Most features in Sweave and the R packages introduced below (except **R2HTML**) are covered by **knitr**, so this section is mainly for historical interest.

The **highlight** package (Francois, 2013) provides syntax highlighting for R code in Rnw documents. Like **pgfSweave**, **cacheSweave**, and **R2HTML** below, **highlight** was extended based on Sweave. In early versions (before v0.6), **knitr** depended on **highlight** to do syntax highlighting, but this dependency was later due to maintenance problems and the fact that it has additional dependencies (the **Rcpp** and the **parser** package). Now **knitr** uses its own syntax highlighting functions, which were based on regular expressions before R 3.0.0 and rely on the function *getParseData()* in the **utils** package in base R after R 3.0.0. To achieve similar functionality as **highlight**, we just need to use the chunk option highlight = TRUE in **knitr**.

The **cacheSweave** package (Peng, 2012) added an important feature to Sweave: the cache system; the **weaver** package (Falcon, 2013) did a similar thing with a different implementation. Chunk options cache and dependson were added, having the same meaning as in **knitr** (see Chapter 8).

The **pgfSweave** package (Bracken and Sharpsteen, 2012) combined the features of **highlight** and **cacheSweave**, and added further support for graphics. Specifically, plots can be cached as well, and TikZ graphics via the **tikzDevice** package are also supported for the sake of font style consistency. The author of this book switched to **pgfSweave** from Sweave when it came out, and contributed the **formatR** support to it (the `tidy` option), but as time went by, it became more and more difficult to keep up with changes in Sweave. This package has been removed from the CRAN repository. At any rate, the design of **knitr** benefited a lot from the author's experience with **pgfSweave**.

The **brew** package (Horner, 2011) is a light-weight templating framework, and its syntax is similar to PHP (`<?php ?>`). Basically it parses and executes R code inside the templating tag `<% %>`. You can think of this as the inline R code in Sweave and **knitr**. It has a cache system but does not have direct graphics support. The **knitr** package also has partial support for the **brew** syntax, which we did not mention in Chapter 5; below is an example that can be compiled through **knitr**:

```
The value of pi is <% pi %>, and 2 times pi is <% 2*pi %>.
```

If an input file has an extension *.brew, **knitr** will use the **brew** syntax automatically. Note **brew** actually supports incomplete code fragments in several inline expressions, which makes it really similar to PHP. Here is an example taken from **brew** but **knitr** will not be able to compile it:

```
<% for (i in c('1+1','1+pi','1+pi','sin(pi/2)')) { -%>
> <%=i%>
<% print(eval(parse(text=i))) %>
<% } -%>
```

The **R2HTML** package (Lecoutre, 2014) contains a large number of functions to export R objects to HTML. The main function is an S3 generic function *HTML()*, which can be applied to a variety of R objects such as data frames, tables, lm objects (returned by *lm()*) and so on. Below is a subset of the `iris` data converted to an HTML table:

```
library(R2HTML)
HTML(head(iris[, -5], 1), "", caption = NULL)



<p align= center >
<table cellspacing=0 border=1><tr><td>
```

```
    <table border=0 class=dataframe>
    <tbody>
    <tr class= firstline >
        <th>   </th>
        <th>Sepal.Length  </th>
        <th>Sepal.Width  </th>
        <th>Petal.Length  </th>
        <th>Petal.Width</th>
    </tr>
<tr>
<td class=firstcolumn>1
</td>
<td class=cellinside>5.1
</td>
<td class=cellinside>3.5
</td>
<td class=cellinside>1.4
</td>
<td class=cellinside>0.2
</td></tr>
    </tbody>
</table>
 </td></table>
```

We can make use of **R2HTML** inside **knitr** for R HTML documents, with the chunk option `results = 'asis'` to write raw HTML code into the output.

The other major contribution of **R2HTML** is the Sweave extension, which allows one to write an HTML report based on Sweave.

There is a task view on CRAN about reproducible research: `http://cran.r-project.org/web/views/ReproducibleResearch.html`, where we can find more packages on this topic.

---

## 16.3    Python Packages

In this section we introduce three packages based on Python for dynamic documents: Dexy, PythonTEX, and IPython.

### 16.3.1  Dexy

Dexy (`http://www.dexy.it`) is a free Python package that features a very general design. According to its website:

> Dexy is a free-form literate documentation tool for writing any kind of technical document incorporating code. Dexy helps you write correct documents, and to easily maintain them over time as your code changes.

The four major features are:

1. any language (source code)

2. any markup (output)

3. any template

4. any API (programming)

There are apparently some similarities between Dexy and **knitr**, such as the multi-language support. An important concept of Dexy is the "filter": the filter takes an input file and converts it to an output file, which is similar to the pipe | in shell scripts. The filters in Dexy are actually a combination of concepts in **knitr**: a filter may render output (e.g., from Markdown to HTML), or run a programming language (like language engines in **knitr**), or do additional tasks like **knitr**'s chunk hooks.

Normally Dexy separates computer code from templates, which can be either good or bad. The good aspect is that the source scripts can be reused, and the bad thing is we have to jump back and forth between the report environment and the source code. By default **knitr** directly embeds code chunks in a report, but we can also externalize code chunks as introduced in Chapter 9.

### 16.3.2  PythonTEX

PythonTEX (`https://github.com/gpoore/pythontex`) is a LATEX package, which features execution of Python code within LATEX. According to its documentation:

> PythonTEX provides fast, user-friendly access to Python from within LATEX. It allows Python code entered within a LATEX document to be executed, and the results to be included within the original document. It also provides syntax highlighting for code within LATEX documents via the Pygments package.

We can insert inline Python code using the \pyb{} command, or emulate a Python session in LaTeX using the pyconsole environment, e.g.,

```
\begin{pyconsole}[][frame=single]
x = 123
y = 345
z = x + y
z
def f(expr):
    return(expr**4)

f(x)
print('Python says hi from the console!')
\end{pyconsole}
```

When we compile this document, the Python code will be evaluated and the results will be inserted into the output.

Due to its Python origin, it also has integration with other Python packages such as SymPy (symbolic manipulation) and matplotlib (plots).

### 16.3.3 IPython

IPython (http://ipython.org) is an interactive shell for Python that features a Web-based notebook with support for code, text, mathematical expressions, inline plots and other rich media, high performance tools for parallel computing, and so on.

Figure 16.1 is a screenshot of IPython in a GNOME terminal under Ubuntu. We can see that it has basic functionalities of a shell such as the auto-completion of commands: we type x.spl<TAB> in the shell and will see the auto-completion below.

The most notable feature related to report generation is its Web-based notebook: we can work in the Web browser with Python commands, view the results on the fly (including both numerical and graphical results), and the notebook can be continuously updated as we input more content into the notebook. It is very much like writing code chunks in **knitr**.

An IPython notebook can be saved as a JSON file with the extension *.ipynb, which can be shared with others. The notebook may or may not contain output; a notebook without the output is similar to the source document for **knitr** (e.g., Rnw and Rmd documents).

Inspired by IPython, **knitr** has got a similar Web notebook (but with fewer features), which we have mentioned in Section 3.2.2.

FIGURE 16.1: A screenshot of IPython: input is marked as In[*n*], and output is marked as Out[*n*].

## 16.4  More Tools

In addition to R and Python packages, there are tools in other programs. It is impossible to enumerate all the tools for dynamic documents in this chapter. Schulte et al. (2012) have provided a list of existing tools for literate programming and reproducible research, such as Javadoc, cweb, noweb, Sweave, SASweave, and so on.

## 16.4.1   Org-mode

Org-mode is a plain text markup language, with an implementation in the Emacs text editor (Schulte et al., 2012). It supports both literate programming and reproducible research (in the sense of dynamic documents). It more or less follows the syntax of early implementations of literate programming such as WEB and noweb, i.e., it has the concept of code chunks and text chunks (the text chunks are sometimes called "prose"). A code chunk in Org-mode looks like this:

```
#+name: c-chunk
#+begin_src C
  int main(){
    return 0;
  }
#+end_src
```

By comparison, the same chunk is written like this in **knitr**:

```
<<c-chunk, engine='c'>>=

int main(){
  return 0;
}

@
```

The metadata is stored in the chunk headers. Org-mode supports any input languages, with either LaTeX or HTML as the output format.

Schulte et al. (2012) mentioned the capability of literate programming of existing tools (e.g., Sweave does not have it), which we did not emphasize in this book because it does not sound interesting to report writers. As a matter of fact, **knitr** also has this capability of reorganizing code chunks (see Chapter 9). Below is a simple example of defining chunk B later but embedding it in an earlier chunk A:

```
<<A>>=
df <- data.frame(x = 1:10, y = rnorm(10))
<<B>>
coef(fit)
@
```

```
<<B>>=
fit <- lm(y ~ x, data = df)
@
```

Powerful as it is, the Emacs nature of Org-mode may be an obstacle to beginners.

### 16.4.2  SASweave

SASweave (http://homepage.cs.uiowa.edu/~rlenth/SASweave) is an implementation of literate programming with SAS and R. It was written in gawk. The basic idea is the same as Sweave and **knitr**. See Lenth and Højsgaard (2007) for more information. The **knitr** package has more comprehensive support for R but less support for SAS compared to SASweave.

### 16.4.3  Office

We do not have to choose the plain text format for dynamic documents, whereas almost everything we have introduced in this book is based on plain text. There are tools based on OpenOffice (or OpenDocument Text) or Microsoft Office products (we call them Office documents for short), and they may seem appealing at first glance. At its core, an Office document is usually an XML file (which may be compressed), so it is possible to embed code chunks in it. We can parse code chunks, run them, and insert the results back.

The major problem we see is that the XML format is too complicated and there are too many standards, so it is not trivial to make sure the modified document is still a valid Office document. As one example, the StatWeave package (http://homepage.stat.uiowa.edu/~rlenth/ StatWeave/) no longer works with OpenOffice (3.2 and higher) because "OpenOffice flags the modified document as corrupted."

By comparison, plain text files are much easier to deal with; there are no complicated standards such as ECMA-376 to take care of. If we want Office documents at all, there are at least possibilities of conversion from Markdown. Recall what we quoted in Chapter 1:

---

The source code is real.

---

# A

## Internals

In this appendix we explain some internal structures of the **knitr** package, which may help other developers better understand this package, and contribute code when necessary. General users do not need to read this appendix. We show the internals in three aspects: documentation, the application of closures, and the implementation of some features.

## A.1 Documentation

There are three types of documentation in **knitr**: the R documentation (Rd), the PDF manuals, and the website.

The R documentation is based on **roxygen2** (Wickham et al., 2015), which allows one to write Rd in roxygen comments (#') with tags, and these comments will be translated into the real Rd. Below is an example of the roxygen comment:

```
#' @author Yihui Xie
```

It will be translated into Rd as:

```
\author{Yihui Xie}
```

There is a series of tags in roxygen such as @usage, @param, @return, and @examples, which correspond to \usage{}, \arguments{\item{}}, \value{}, and \examples{}, respectively, in Rd. The advantage of writing roxygen comments over the official Rd is that we can keep the documentation and the source code in the same file; by comparison, the official approach to writing R packages is to write R sources under the R/ directory, and manual pages as *.Rd files under man/. This is not convenient because we have to jump between two files, and it is likely that we update the R source but forget to update the documentation. Roxygen comments appear right above the R functions in the source, so it is much easier to maintain both the source and documentation.

Below is a complete example of a function documented with roxygen comments:

```
#' Repeat a character string
#'
#' Repeat a string n times and make one string.
#' @param x a character string
#' @param n an integer
#' @return A character string.
#' @examples f('hi', n = 5)
f <- function(x, n = 10) {
    paste(rep(x, n), collapse = "")
}
```

We can use the *roxygenize()* function in **roxygen2** to convert roxygen comments to the official Rd files. All objects in **knitr** are documented in this way. Besides, **roxygen2** also handles NAMESPACE and the `Collate` field in DESCRIPTION automatically, so we can really focus on working R source files.

The source documents of the PDF manuals are under the examples directory (see inst/examples/ in the source package), e.g., the main manual is knitr-manual.Rnw. The Rnw files are exported from LyX files (Section 4.2), so it is recommended to open the LyX files to edit or compile PDF manuals. The PDF manuals are not shipped with the source package, because (1) I do not want to put binary files under version control (especially when they are by-products of source files) and (2) they are hosted in the package website.

The package website is built on Jekyll as introduced in Section 13.4. Specifically, all pages are written in Markdown, and put under the gh-pages branch in the Git repository (the package itself is in the master branch). Github will rebuild the website automatically once changes are pushed there through Git. If you want to contribute to the website, just switch to the gh-pages branch, and update the Markdown files.

## A.2 Closures

Closures play a central role in **knitr**; some common objects such as opts_chunk (Section 5.1.1) and knit_engines (Chapter 11) are built on closures.

A closure is essentially a function, and it also has access to non-local variables. Below is a simple example:

```
f <- function() {
    x <- 1
    function(y) x + y
}
g <- f()
g(5)    # add 5 to x
```

```
## [1] 6
```

```
ls(environment(g))    # g can see x
```

```
## [1] "x"
```

The function *g()* was created from *f()* (note *f()* returns a function), *g()* uses an object x that was created inside *f()*, and x only exists in *f()*. No matter where *g()* is called, it always has access to this x.

In fact, we can even modify non-local variables through a closure. Below is a minimal example that shows how the chunk options manager `opts_chunk` works:

```
new_list <- function(default = list()) {
    list(get = function() default, set = function(...) {
        x <- list(...)
        if (length(x)) default[names(x)] <<- x
    })
}
```

The function *new_list()* returns a list of functions (a setter and a getter). The object `default` is bound to these two functions. You can think of it as the default list of chunk options. Next we show how to get and set the chunk options.

```
opts <- new_list(list(eval = TRUE))
str(opts$get())
```

```
## List of 1
##  $ eval: logi TRUE
```

```
opts$set(eval = FALSE)    # change eval to FALSE
opts$set(results = "markup")    # add a chunk option
str(opts$get())
```

```
## List of 2
##  $ eval   : logi FALSE
##  $ results: chr "markup"
```

```
opts$set(results = "hide")   # change the results option
```

In the *$set()* function, we used `<<-` to assign the arguments to the object `default`, and that is why we can modify this object in the parent environment (had we used the normal `<-`, `default` in the parent environment would not be modified; a local copy will be created instead).

By using closures, **knitr** can manage objects in their own environments with the same syntax. The internal function *new_defaults()* in **knitr** is used to create such a list of closures.

Besides the objects `opts_chunk` (for managing chunk options) and `knit_engines` (for managing language engines), there are a few other similar objects:

**opts_knit**  package options (Section 12.2)

**opts_current**  chunk options for the current chunk

**opts_template**  chunk option templates (Section 12.1.2)

**knit_hooks**  hook functions (both output hooks and chunk hooks)

**knit_patterns**  syntax patterns for the parser (Section 5.1)

## A.3  Implementation

This section explains some implementation details for this package. One minor thing to mention first is that I use = instead of `<-` as the assignment operator, and you will see = all over the place in the source code. It is a matter of personal taste, and I do not see real disadvantages in it, but you are expected to follow = when contributing code to this package. In this book, you see `<-` because I typed equal signs but they were automatically replaced by **formatR**.

### A.3.1  Parser

The document parser (Section 5.1) works like this: the child elements `chunk.begin` and `chunk.end` in the syntax pattern object are used to split the document into pieces (code chunks and text chunks), and for the code chunks, the chunk options (i.e., the text extracted from the first line) are parsed as R code, and this is why chunk options have to follow the R syntax. Here is an example explaining how **knitr** gets chunk options from a text fragment:

```
## suppose this is the chunk options text
txt <- "label, eval=TRUE, echo=1:3, foo=if(TRUE) 2 else 5"
opc <- eval(parse(text = paste("alist(", txt, ")")))
names(opc)  # the chunk label is not named

## [1] ""     "eval" "echo" "foo"

str(opc)  # some are unevaluated expressions

## List of 4
##  $      : symbol label
##  $ eval: logi TRUE
##  $ echo: language 1:3
##  $ foo : language if (TRUE) 2 else 5
```

First we added the function *alist()* around the text, and this function will treat its arguments as if they described function arguments, therefore no "arguments" will be evaluated at this time. However, the syntax must be valid at least; one exception is the chunk label: it is automatically quoted if necessary, since it is supposed to be a character string. The internal function *parse_params()* is used to parse chunk options:

```
p <- knitr:::parse_params
str(p("chunk-label, eval=TRUE, foo=5"))

## List of 3
##  $ label: chr "chunk-label"
##  $ eval : logi TRUE
##  $ foo  : num 5

# 2a is not a valid symbol in R, but knitr will quote it
# automatically so parsing is OK
parse(text = "alist(2a)")

## Error:  <text>:1:8:  unexpected symbol
## 1:  alist(2a
##               ^

str(p("2a, eval=FALSE"))

## List of 2
##  $ label: chr "2a"
##  $ eval : logi FALSE

str(p("'2a', eval=FALSE"))  # or you can quote it manually
```

```
## List of 2
##  $ label: chr "2a"
##  $ eval : logi FALSE
```

The chunk options are not evaluated until before the chunks are executed, so the chunk options can use objects of unknown values in the document at the parsing time. For example, the options echo and foo above are unevaluated expressions, and we will evaluate them explicitly later:

```
eval(opc$echo)
```

```
## [1] 1 2 3
```

```
eval(opc$foo)
```

```
## [1] 2
```

All code chunks are stored as a named list in an internal object knit_code; the names are chunk labels, and the content is the code. This object is also created as a list of closures, so it has the *get()* and *set()* methods, but it is not recommended to modify this object due to possible unexpected consequences. If needed, we can access code chunks via knitr:::knit_code$get('chunk-label').

## A.3.2   Chunk Hooks

There is a number of default hooks in knit_hooks, which are output hooks (Section 5.3):

```
names(knit_hooks$get(default = TRUE))
```

```
##  [1] "source"    "output"    "warning"   "message"
##  [5] "error"     "plot"      "inline"    "chunk"
##  [9] "text"      "document"
```

Any other hooks in this object are treated as chunk hooks (Chapter 10). Before and after a code chunk is executed, all extra hooks will be called. Here is the pseudo code:

```
hook(before = TRUE, ...)
evaluate(code)
hook(before = FALSE, ...)
```

One issue to keep in mind is the order of the hooks to run: if there are two hooks A and B defined in knit_hooks, what is the order in which they are called? This order is obtained from chunk options: there must be two chunk options, A and B, corresponding to these two hooks, and the order of chunk options determines the order in which to run the hooks; e.g., if A is before B, then hook A is called before B. However, after a code chunk has been evaluated, the order is reversed, and the reason is to make sure the results returned by the hooks pair in groups. For example, suppose the hook A returns \begin{Aenvir} before a chunk, and \end{Aenvir} after a chunk; similarly B returns Benvir. Then what we want in the output is this:

```
\begin{Aenvir}
\begin{Benvir}
% results from the chunk
\end{Benvir}
\end{Aenvir}
```

Note \end{Benvir} comes before \end{Aenvir}. For this reason, the following two chunks return different results when hooks A and B are defined:

```
<<A=TRUE, B=TRUE>>=
<<B=TRUE, A=TRUE>>=
```

### A.3.3   Option Aliases

It takes only a few lines to implement chunk option aliases (Section 12.1.1), since it is a simple operation of substituting certain elements in a list. Below is a short function that illustrates the idea:

```
apply_aliases <- function(x, list) {
    ## names are aliases of x
    list[x] <- list[names(x)]
    list
}
al <- c(w = "fig.width", h = "fig.height", a = "fig.align")
op <- list(w = 7, h = 7, echo = TRUE, a = "center")
str(op)  # user's options

## List of 4
##  $ w   : num 7
```

```
## $ h   : num 7
## $ echo: logi TRUE
## $ a   : chr "center"

str(apply_aliases(al, op))  # corrected options

## List of 7
## $ w         : num 7
## $ h         : num 7
## $ echo      : logi TRUE
## $ a         : chr "center"
## $ fig.width : num 7
## $ fig.height: num 7
## $ fig.align : chr "center"
```

Aliases are set in a named character vector, and the names are the aliases of the elements in the vector. In the above example, *apply_aliases()* added elements fig.width and fig.height into the list op according to the values of w and h, respectively, which were specified by the user, but internally **knitr** still uses fig.width and fig.height.

### A.3.4   Cache

The cache in **knitr** is also managed by an object consisting of closures, but it is more complicated (see the internal function *new_cache()*). The closures are used to save, load, and delete cache files, and we only explain one aspect of the cache here: how the side effect of printing is cached (Section 8.4).

As we mentioned in Section 5.3, the code chunks are evaluated by the **evaluate** package. As a matter of fact, printed results are returned as character strings, and the output of the whole chunk is also a character string (formatted by output renderers). This character string is assigned to a variable, with the variable name constructed from the MD5 hash and the chunk label. This variable is saved in the cache database along with all other variables created in the chunk. The next time the chunk is to be evaluated, **knitr** will check if the chunk needs to be updated; if not, all objects will be loaded directly, including the object of the chunk output, which also contains the printed results (in fact, everything of this chunk); instead of re-evaluating the chunk, this object is written into the output directly.

### A.3.5 Compatibility with Sweave

Since **knitr** uses some different chunk options with Sweave, there is a function *Sweave2knitr()* to correct the inappropriate options and their values. For example, `results = tex` is changed to `results = 'markup'` automatically (because `'tex'` is not an appropriate value to reflect what the `results` option really does).

The implementation is mainly based on regular expressions, and here is a simple example:

```
op <- "<<eval=TRUE, results=tex>>="
gsub("(results)\\s*=\\s*tex", "\\1='markup'", op)

## [1] "<<eval=TRUE, results='markup'>>="
```

*Sweave2knitr()* takes care of a large number of cases of inappropriate chunk options as well as \SweaveOpts{} and \SweaveInput{}. See Section 16.1 for examples.

### A.3.6 Concordance

The concept of concordance is specific to Rnw/LATEX. The problem to solve is the mapping of line numbers between the TEX output and the Rnw source. When an error occurs in LATEX, we know the line number of the problematic line (by parsing the error log), but we do not know the corresponding line number in the Rnw source document, because the line numbers of the two documents may not match. One chunk of 5 lines in the Rnw document may produce 10 or 3 lines of LATEX code in the output.

Sweave has a better implementation of concordance than **knitr**. The mapping is more precise in Sweave. In **knitr**, it is only an approximation achieved in this way: when parsing the source document, the number of lines of the code chunks and text chunks are recorded; after these chunks have been evaluated, the number of lines of the corresponding output chunks is calculated again. Suppose one source chunk has 5 lines, and if

- the output has 5 lines too, the $i$-th line in the source is mapped to the $i$-th line in the output

- the output has 3 lines, the first 3 lines of the source are mapped to the 3 lines in the output

- the output has 10 lines, the 5 lines of the source are mapped to the first 5 lines in the output

Obviously this may not be a good approximation, but it should be helpful enough for error navigation. At least the error number in LaTeX can point to a rough area of the problematic source.

The other use of concordance is the navigation between PDF and Rnw files. SyncTeX supports this kind of navigation: you can click one line in the PDF document to jump back to the source file, or click one line in the source to jump to the PDF. Without the concordance information, we cannot navigate between Rnw and PDF (only TeX↔PDF is possible).

For now, only RStudio uses the concordance information produced by **knitr**. To enable concordance (it is disabled by default), you can set the package option (RStudio does this automatically):

```
opts_knit$set(concordance = TRUE)
```

When concordance is enabled, a file input-concordance.tex will be generated if the Rnw file is named as input.Rnw. This file contains compressed mapping information.

## A.4   Syntax

Users may wonder why **knitr** uses different input syntax for different document formats (Section 5.1), e.g., Rnw uses <<>>=, and Rmd uses ```` ```{r}````. In fact, the syntax is not tied to document formats; we can certainly use the Rnw syntax for Rmd documents.

```
# This is a markdown document

Here is a **code chunk**:

<<test>>=
1 + 1
rnorm(5)
@

And an inline value \Sexpr{pi}.
```

For the example document above (suppose it is named test.Rmd), we can compile it by:

```
library(knitr)
pat_rnw()          # input is Rnw syntax
render_markdown()  # output is markdown
knit("test.Rmd")
```

The function *pat_rnw()* sets the syntax to be Rnw, and the function *render_markdown()* sets the output renders to be Markdown hooks.

But why not use the Rnw syntax for all documents? The decision was made because I wanted more natural syntax according to the authoring format, and <<>>= is not a valid markup in any document format; e.g., it is neither a LaTeX command nor an HTML tag. In fact, Sweave has another set of syntax that is LaTeX-like, e.g.,

```
\begin{Scode}{fig = TRUE, echo = FALSE}
library("graphics")
boxplot(Ozone ~ Month, data = airquality)
\end{Scode}
```

I would prefer [] to {} for chunk options, which will be a more natural choice in LaTeX. Anyway, <<>>= remained in **knitr** due to its popularity.

Except for Rnw documents (due to historic reasons), other formats make the **knitr** source documents still valid documents even before the R code is executed. For example, R code in R HTML documents is put in HTML comments (<!-- -->).

# Bibliography

Adler, D. and Murdoch, D. (2014). *rgl: 3D visualization device system (OpenGL)*. R package version 0.95.1201.

Allaire, J., Cheng, J., Xie, Y., McPherson, J., Chang, W., Allen, J., Wickham, H., and Hyndman, R. (2015a). *rmarkdown: Dynamic Documents for R*. R package version 0.5.1.

Allaire, J., Horner, J., Marti, V., and Porte, N. (2015b). *markdown: Markdown Rendering for R*. R package version 0.7.7.

Auguie, B. (2013). *cda: Coupled dipole approximation in electromagnetic scattering*. R package version 1.3.3.

Baggerly, K. A., Morris, J. S., and Coombes, K. R. (2004). Reproducibility of seldi-tof protein patterns in serum: comparing datasets from different experiments. *Bioinformatics*, 20(5):777–785.

Bracken, C. and Sharpsteen, C. (2012). *pgfSweave: Quality speedy graphics compilation and caching with Sweave*. R package version 1.3.0.

Buckheit, J. and Donoho, D. (1995). Wavelab and reproducible research. *Wavelets and Statistics*, 103:55.

Chang, W., Cheng, J., Allaire, J., Xie, Y., and McPherson, J. (2015). *shiny: Web Application Framework for R*. R package version 0.11.1.

Dahl, D. B. (2014). *xtable: Export tables to LaTeX or HTML*. R package version 1.7-4.

Eddelbuettel, D., Francois, R., Allaire, J., Ushey, K., Bates, D., and Chambers, J. (2015). *Rcpp: Seamless R and C++ Integration*. R package version 0.11.5.

Ellson, J., Gansner, E., Koutsofios, L., North, S., and Woodhull, G. (2002). Graphviz — open source graph drawing tools. In *Graph Drawing*, pages 483–484. Springer-Verlag.

Falcon, S. (2013). *weaver: Tools and extensions for processing Sweave documents*. R package version 1.26.0.

Fomel, S. and Claerbout, J. (2009). Guest editors' introduction: Reproducible research. *Computing in Science & Engineering*, 11(1):5–7.

Francois, R. (2013). *highlight: Syntax highlighter*. R package version 0.4.3.

Friedl, J. (2006). *Mastering Regular Expressions*. O'Reilly Media, Incorporated.

Gentleman, R. (2005). Reproducible research: A bioinformatics case study. *Statistical Applications in Genetics and Molecular Biology*, 4(1):1034.

Gentleman, R. and Temple Lang, D. (2004). Statistical analyses and reproducible research. *Bioconductor Project Working Papers*. URL: http://biostats.bepress.com/bioconductor/paper2.

Gove, J. H. (2013). *sampSurf: Sampling Surface Simulation for Areal Sampling Methods*. R package version 0.6-8.

Gruber, J. (2004). *The Markdown Project*. URL: http://daringfireball.net/projects/markdown/.

Guo, J., Betancourt, M., Brubaker, M., Carpenter, B., Gao, Y., Goodrich, B., Hoffman, M., Lee, D., Li, P., Malecki, M., and Gelman, A. (2014). *rstan: RStan: R interface to Stan*. R package version 2.5.0.

Harrell, Jr., F. E. (2001). *Regression Modeling Strategies: With Applications to Linear Models, Logistic Regression, and Survival Analysis*. Springer New York.

Harrell, Jr., F. E. (2015). *Hmisc: Harrell Miscellaneous*. R package version 3.15-0.

Horner, J. (2011). *brew: Templating Framework for Report Generation*. R package version 1.0-6.

Horton, N., Aloisio, K., Zhang, R., and Loi, L. (2012). The statistical sleuth (2nd edition) in R. URL: http://www.math.smith.edu/~nhorton/sleuth/.

Huang, Y. and Gottardo, R. (2013). Comparability and reproducibility of biomedical data. *Briefings in Bioinformatics*, 14(4):391–401.

Ihaka, R. and Gentleman, R. (1996). R: A language for data analysis and graphics. *Journal of Computational and Graphical Statistics*, 5(3):299–314.

Jockers, M. L. (2014). *Text Analysis with R for Students of Literature*. Springer.

Knuth, D. E. (1983). The WEB system of structured documentation. Technical report, Department of Computer Science, Stanford University.

Knuth, D. E. (1984). Literate programming. *The Computer Journal*, 27(2):97–111.

Lebanon, G. (2012). *Probability: The Analysis of Data*, volume 1. CreateSpace Independent Publishing Platform.

Lecoutre, E. (2014). *R2HTML: HTML exportation for R objects*. R package version 2.3.1.

Leisch, F. (2002). Sweave: Dynamic generation of statistical reports using literate data analysis. In *COMPSTAT 2002 Proceedings in Computational Statistics*, number 69, pages 575–580. Heidelberg: Physica Verlag.

Lenth, R. V. and Højsgaard, S. (2007). Sasweave: Literate programming using sas. *Journal of Statistical Software*, 19(8):1–20.

Murdoch, D. (2012). *tables: Formula-driven table generation*. R package version 0.7.

Murphy, D. (2012). Changes and additions to **ggplot2** 0.9.0. URL: https://github.com/djmurphy420/ggplot2-transition-guide.

Murrell, P. (2011). *R Graphics, Second Edition*. Chapman & Hall/CRC.

Murrell, P. and Ripley, B. (2006). Non-standard fonts in PostScript and PDF graphics. *R News*, 6(2):41–47.

Oetiker, T., Partl, H., Hyna, I., and Schlegl, E. (1995). *The not so short introduction to LATEX2ε*. URL: http://www.ctan.org/tex-archive/info/lshort/.

Peng, R. (2009). Reproducible research and biostatistics. *Biostatistics*, 10(3):405–408.

Peng, R. D. (2012). *cacheSweave: Tools for caching Sweave computations*. R package version 0.6-1.

Qiu, Y. and Xie, Y. (2015). *highr: Syntax Highlighting for R Source Code*. R package version 0.5.

Qiu, Y., Xie, Y., and Bracken, C. (2015). *R2SWF: Convert R Graphics to Flash Animations*. R package version 0.9.

R Core Team (2014). *R Language Definition*. R Foundation for Statistical Computing, Vienna, Austria.

R Core Team (2015). *R: A Language and Environment for Statistical Computing*. R Foundation for Statistical Computing, Vienna, Austria.

Ramsey, F. and Schafer, D. (2002). *The Statistical Sleuth: A Course in Methods of Data Analysis, Second Edition*. Duxbury Press.

Ramsey, N. (1994). Literate programming simplified. *Software, IEEE*, 11(5):97–105.

Rossini, A. (2002). Literate statistical analysis. In *Proceedings of the 2nd International Workshop on Distributed Statistical Computing*, pages 15–17, Vienna, Austria.

Rossini, A., Heiberger, R., Sparapani, R., Maechler, M., and Hornik, K. (2004). Emacs speaks statistics: A multiplatform, multipackage development environment for statistical analysis. *Journal of Computational and Graphical Statistics*, 13(1):247–261.

Schulte, E., Davison, D., Dye, T., and Dominik, C. (2012). A multilanguage computing environment for literate programming and reproducible research. *Journal of Statistical Software*, 46(3):1–24.

Sharpsteen, C. and Bracken, C. (2015). *tikzDevice: R Graphics Output in LaTeX Format*. R package version 0.8.1.

Tantau, T. (2008). *The TikZ and PGF Packages*. URL: http://sourceforge.net/projects/pgf/.

Tantau, T., Wright, J., and Miletic, V. (2012). *User's Guide to the Beamer Class*. URL: http://bitbucket.org/rivanvx/beamer.

Temple Lang, D., Swayne, D., Wickham, H., and Lawrence, M. (2014). *rggobi: Interface between R and GGobi*. R package version 2.1.20.

Vaidyanathan, R. (2012). *slidify: Generate reproducible html5 slides from R markdown*. R package version 0.4.5.

Vaidyanathan, R., Cheng, J., Allaire, J., Xie, Y., and Russell, K. (2014). *htmlwidgets: HTML Widgets for R*. R package version 0.3.2.

van Heesch, D. (2008). Doxygen: Source code documentation generator tool. URL: http://www.doxygen.org/.

Venables, W. N. and Ripley, B. D. (2002). *Modern Applied Statistics with S*. Springer-Verlag, 4th edition.

Wei, T. (2013). *corrplot: Visualization of a correlation matrix.* R package version 0.73.

Wickham, H. (2015). *evaluate: Parsing and Evaluation Tools that Provide More Details than the Default.* R package version 0.7.

Wickham, H., Danenberg, P., and Eugster, M. (2015). *roxygen2: In-Source Documentation for R.* R package version 4.1.1.

Xie, Y. (2013). *runr: Run External Programs from R.* R package version 0.0.6.

Xie, Y. (2014). *printr: Automatically Print R Objects According to knitr Output Format.* R package version 0.0.3.

Xie, Y. (2015a). *formatR: Format R Code Automatically.* R package version 1.2.

Xie, Y. (2015b). *knitr: A General-Purpose Package for Dynamic Report Generation in R.* R package version 1.10.

Xie, Y. (2015c). *servr: A Simple HTTP Server to Serve Static Files or Dynamic Documents.* R package version 0.2.

Yin, T., Cook, D., and Lawrence, M. (2012). ggbio: an R package for extending the grammar of graphics for genomic data. *Genome Biology,* 13(8):R77.

# *Index*

animation, 70
AsciiDoc, 36
awk, 117, 122

bash, 111, 112, 146
Beamer, 135, 194

C, 115
C++, 113
cache, 81, 150, 254
child documents, 95
chunk dependency, 86
chunk headers, 28
chunk hooks, 99, 252
chunk label, 29, 79
chunk options, 13, 28, 49, 101, 127
chunk reference, 91, 130
code chunks, 12
code externalization, 93
CoffeeScript, 117

DataTables, 209
debugging, 160
dynamic documents, xiii, 2

Emacs, 25
encoding, 62
EPUB, 207
ESS, 2, 25

figure alignment, 71
figure size, 72
Fortran, 115

Github, 11, 166, 220
graphical device, 60

Graphviz, 121
Groovy, 117

Haskell, 117
Highlight, 122
HTML, 15, 33, 35, 138, 144, 151,
    173, 220, 221, 224
HTML widgets, 208
HTML5, 191

image URI, 144, 151, 162
inline R code, 13, 138
interactive documents, 199
ioslides, 191

JavaScript, 208
Jekyll, 165, 219

knit(), 12, 26, 146
knit_expand(), 155
knit_rd(), 151, 220

language engine, 111
LaTeX, 12, 23, 35, 132, 138, 184
listings, 41, 133
literate programming, 1, 30, 44,
    230
LyX, 23, 227

Makefile, 226
Markdown, 15, 31, 144, 163, 165,
    219

Node.js, 117
notebook, 15, 23
Noweb, 6